T0248724

# Guide to
# Metering Systems

# Publication Information

Published by The Institution of Engineering and Technology, London, United Kingdom

The Institution of Engineering and Technology is registered as a Charity in England & Wales (no. 211014) and Scotland (no. SC038698).

© The Institution of Engineering and Technology 2017

First published 2016

This publication is copyright under the Berne Convention and the Universal Copyright Convention. All rights reserved. Apart from any fair dealing for the purposes of research or private study, or criticism or review, as permitted under the Copyright, Designs and Patents Act, 1988, this publication may be reproduced, stored or transmitted, in any form or by any means, only with the prior permission in writing of the publishers, or in the case of reprographic reproduction in accordance with the terms of licences issued by the Copyright Licensing Agency. Enquiries concerning reproduction outside those terms should be sent to the publisher at this address:

The Institution of Engineering and Technology,
Michael Faraday House,
Six Hills Way, Stevenage,
SG1 2AY, United Kingdom.

Copies of this publication may be obtained from:

The Institution of Engineering and Technology
PO Box 96, Stevenage, SG1 2SD, UK
Tel: +44 (0)1438 767328
Email: sales@theiet.org
www.electrical.theiet.org/books

While the publisher and contributors believe that the information and guidance given in this work is correct, all parties must rely upon their own skill and judgement when making use of it. Neither the publisher nor contributors assume any liability to anyone for any loss or damage caused by any error or omission in the work, whether such error or omission is the result of negligence or any other cause. Any and all such liability is disclaimed.

The moral rights of the authors to be identified as author of this work have been asserted by them in accordance with the Copyright, Designs and Patents Act 1988.

A list of organisations represented on this committee can be obtained on request to IET Standards. This publication does not purport to include all the necessary provisions of a contract. Users are responsible for its correct application. Compliance with the contents of this document cannot confer immunity from legal obligations.

It is the constant aim of the IET to improve the quality of our products and services. We should be grateful if anyone finding an inaccuracy or ambiguity while using this document would inform the IET Standards development team at IETStandardsStaff@theiet.org or the IET, Six Hills Way, Stevenage SG1 2AY, UK.

**ISBN 978-1-78561-059-2** (paperback)
**ISBN 978-1-78561-060-8** (electronic)

# CONTENTS

© The Institution of Engineering and Technology

# Table of Figures

# Acknowledgements

The IET would like to thanks the following parties for their contributions to this document:

*Lead author:*

Vic Tuffen, Tuffentech Services Ltd.

*Working group contributors:*

Bill Wright (Chair), Electrical Contractors' Association
John Cowburn, Smart Energy Networks/UK Metering Forum
Andy Baker, Association of Meter Operators
Huw Blackwell, Islington Borough Council
Leighton Burgess, Regulatory Delivery, part of the Department for Business, Energy and Industrial Strategy (BEIS)
Hywel Davies, CIBSE
Niall Enright, Sustain success
Andy Godley, WRc plc
Andy Lewry, BRE
Dale Meadows, COFELY GDF Suez
David Moorhouse, Regulatory Delivery, part of the Department for Business, Energy and Industrial Strategy (BEIS)
Peter Morgan, Department for Business, Energy and Industrial Strategy (BEIS)
Nigel Orchard, Pilot Systems
Sara Shaw, CBRE
Mark Simmons, Switch-2 Energy Ltd (formerly known as ENER-G Switch 2 Ltd)
Cameron Steel, BK Design Associates UK Ltd
Kris Szajdzicki, ND Metering Ltd
Brian Taylor, Matrix Control Solutions Ltd (a company of e.on)
Peter Tse, BSRIA

© The Institution of Engineering and Technology

# SECTION 1

# Introduction

## 1.1 Executive summary

It is a well understood maxim that to manage something you must first be able to measure it.

Whilst meters by themselves will provide useful data, when they are combined in a system for collecting, managing and presenting the data in a useful way they can become essential and powerful tools.

One of the main drivers for the use of metering systems is to meet a regulatory need. The legal requirements that govern the way primary fiscal meters are managed is well established but with the advent of legislation relating to carbon reduction and energy saving, the business drivers for sub-metering systems have moved beyond being able to allocate costs or identifying wastage.

This Guide provides a basic introduction to the field of secondary or sub-metering in four key areas: electricity, gas, water and heat. It will guide the reader through the steps required in order to make decisions on where to use metering systems and how to adopt them. It will also look at the key technology areas in turn and provide some guidance on the selection and installation and commissioning of each one.

The Guide has been written with due regard to other publications in this area, most notably:

**(a)** Carbon Trust *Metering technology overview* (CTV027);
**(b)** Carbon Trust *Monitoring and targeting* – in depth management guide (CTG008);
**(c)** CIBSE *Building Energy Metering* (TM39);
**(d)** CIBSE *Energy Benchmarks* (TM46);
**(e)** DECC *Enhanced Capital Allowance Energy Technology Criteria Lists* (2009/2013);
**(f)** Department for Communities and Local Government Building Regulations; and
**(g)** BPF Office ABI *Landlord's Energy Statement Guidance and Specification* Version 2.3 April 2007.

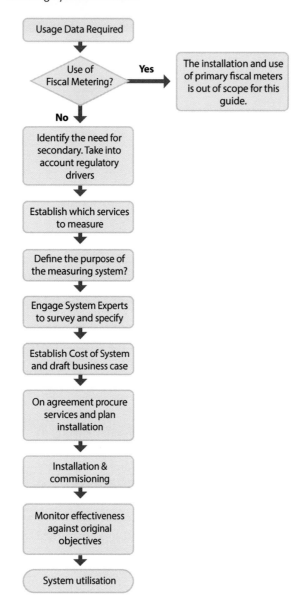

## 1.2 Scope

This Guide covers the regulatory and technical requirements of metering systems used in residential, industrial, commercial and renewable energy markets. It provides guidance on defining the purpose, installation and use of those metering systems.

**Note:**    References to electricity, gas and water utility metering are out of the scope of this Guide, but should be noted as an additional information source.

The purpose of this Guide is to provide guidance on:

**(a)** the safe and effective approaches to achieving compliant metering systems that support cost savings, energy management and the establishment of resource efficiency measures.

**(b)** the technical specification, design, selection, installation, commissioning, operation and maintenance of metering, including the integration of metering into organisational management systems.

**(c)** all types of metering that are relevant to residential/industrial/commercial/ renewable energy applications.

© The Institution of Engineering and Technology

# 1.3 Role of metering

There are increasing requirements for metering systems to be integrated into energy management systems so that targeted reporting can be achieved and data can be provided for energy management systems or other facilities within a building. However, this entails compatible data logging and communications networking to ensure that regular and accurate meter readings can be taken without the cost of manual readings.

Metering systems are essential for:

**(a)** managing energy and environment services;
**(b)** cost allocation; and
**(c)** resource efficiency of systems and subsystems in industrial and commercial applications.

Good practice when applying metering is fundamental to successful regulatory compliance, operational and resource efficiency and cost saving. Good practice entails the appropriate design, specification, installation and integration of metering systems, as well as suitable management oversight and responsibility.

Secondary or sub-metering is the implementation of individual electricity, gas, water or other specialist meters that allow utility usage to be individually measured. This enables asset managers and energy providers to identify energy usage of individual buildings, plants and equipment for financial and energy management purposes.

Key benefits of metering and monitoring include:

**(a)** cost visibility and, as a result, financial verification;
**(b)** optimisation of equipment performance, system monitoring or condition-based maintenance;
**(c)** maximised energy savings and integration of on-site energy systems through monitoring and targeting;
**(d)** enhanced load control and targeted demand response peak shaving measures; and
**(e)** improved building design and use of assets.

## 1.3.1 Metering – regulations and standards

Fiscal metering used for billing purposes is already widely regulated and standardized. The requirement to use meters as a basis for charging consumers is covered by the relevant electricity and gas acts of parliament. In addition, where companies wish to use meters to charge for water usage, this is covered by the Water Industry Act.

The standards that fiscal meters are manufactured to for use by domestic and small and medium-size enterprises (SMEs) are defined in the Measuring Instruments Directive (MID). Meters for very large applications are covered by ELEXON (http://www.elexon.co.uk/) codes of practice in the electricity market and by individual contracts in the gas market.

Additional requirements for smart metering are covered by Smart Metering Equipment Technical Specifications (SMETS) from the UK government's Energy Department, which is currently part of the Department of Business, Energy & Industrial Strategy (BEIS) specifications. Installation and management codes are underpinned by licence conditions, which govern how energy suppliers operate.

| Use of meter | Utility meter | Sub-meter |
|---|---|---|
| Billing | Not covered. | Covered. |
|  |  | Approved meters only. |
| Monitoring | For information purposes only. | Covered. |

## 1.4  Aims of this Guide

The aim of this Guide is to provide an appropriate process for specifying and implementing metering systems that are suited to an organisation's key services or systems, where metering may be applied as either part of an energy/environmental management system or as a regulatory/compliance requirement.

The Guide details the key steps to take when applying metering systems, from design and specification, through the technology selection, installation and commissioning stages, and finally to operation and maintenance.

Checklists for metering key services (such as electricity, gas, water, and heat) are also provided to support organisational policies, practices and procedures.

Key benefits of the Guide:

**(a)** it outlines the role of metering within energy and environment management systems and resource efficiency strategies.
**(b)** it highlights key process stages in the application of fit-for-purpose metering within the context of overall management systems.
**(c)** it provides insight into the end use of meter data and the key technical steps to enable safe and effective metering systems.

## 1.5  Applications of this Guide

This Guide is written for those who need to use/specify metering in order to manage energy/cost savings, for example:

**(a)** energy managers;
**(b)** environment and sustainability managers;
**(c)** engineering and process managers;
**(d)** facilities managers;
**(e)** building owners; and
**(f)** operators.

This Guide can be used:

**(a)** to specify and implement metering systems to client requirements, i.e. electrical contractors and building technicians, mechanical and engineering (M & E) consultants, and commissioning and building service engineers.
**(b)** as a reference for metering market development and regulatory compliance, i.e. meter suppliers, manufacturers and providers, property developers and other end users.

© The Institution of Engineering and Technology

**(c)** for commercial purposes, such as in offices, public buildings, private residential sites.

**(d)** in campus environments, for example, public spaces/sites, universities, industrial parks.

**(e)** for industrial purposes, for example, production plant/processes, laboratories, workshops.

**(f)** for energy management/generation, for example, end-use of fuel, renewable energy import/export, specific energy efficiency monitoring measures.

© The Institution of Engineering and Technology

# SECTION 2

# Considerations for the specification, installation and use of metering equipment

## 2.1 Establishing the need for secondary/ sub-metering

### 2.1.1 General

Meters do not solve problems on their own. In fact the capital investment involved and, in some cases, the practical problems of installing them could, at first sight, present a barrier to using them, but they can be useful tools that help to address particular issues. In some cases they can be the only means to help resolve a problem.

In this section we will look at the thought process and deliberations that are required to make a decision on when and where to use secondary metering, as illustrated in Figure 2.1.

▼ **Figure 2.1** Considerations prior to installation of a sub-metering system

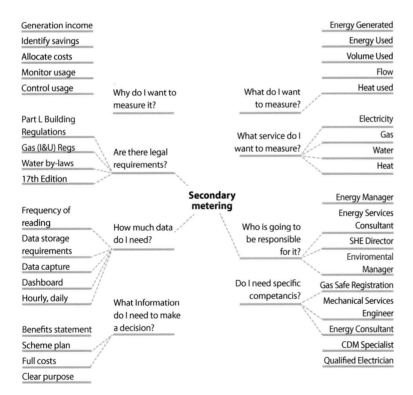

In terms of business planning for the targeted application of metering systems and the use of meter data, the following hierarchy applies.

▼ **Table 2.1**    Hierarchy of decision maker

| Role | Rationale/driver | Key questions |
|---|---|---|
| Director/board level | Compliance with organisational strategy | Why is metering important? |
| Energy/facilities/site manager level | Carbon/energy/cost/ management | What are the key considerations for metering? |
| Operations/maintenance technician level | Performance/effectiveness of key sites/buildings/ equipment | How should metering be implemented for key services? |

## 2.1.2   Key questions in establishing sub-metering

Within this hierarchy, the key questions that energy facilities or site managers need to ask when looking to improve resource efficiency are:

**(a)** what needs to be managed? (Why is metering important at a policy/strategy level?)

**(b)** what needs to be measured? (What metering criteria/data/monitoring are required?)

**(c)** how are metering systems evaluated as being fit for purpose? (How are systems installed, commissioned and integrated?)

**(d)** who manages the installation/benefits delivery? (Who is responsible for monitoring/targeting energy use?)

**(e)** who funds the meter installations?

## 2.1.3   Example: application of questions in a sub-metering scenario

The manager of a large industrial site is responsible for its manufacturing facilities, power and heat requirements, offices, staff welfare facilities including a kitchen and canteen and a new gas-fired combined heat and power plant.

Some of the units on the site are rented out to other group companies and independent traders who are using electricity, gas and water. The business itself spends a great deal of money on electricity, gas and water, which includes heating water for the manufacturing process.

In order for the directors to know what the money is being spent on and whether the bills can be reduced, they must first determine what utilities each part of the business is using.

### Step 1: understanding how to find this information

The first step towards establishing what sort of metering is required is to first determine what needs to be measured. At first sight this may appear to be straightforward. An office, for example, is going to use electricity for lighting and computers. It may have an air conditioning load for the office space and a server room. Each of the circuits can be measured independently. If the offices are heated with a gas boiler and the boiler heats only that business it may be a simple job of installing a check meter to measure the gas used by that boiler.

© The Institution of Engineering and Technology

However, if the boiler serves more than one business or if waste heat from the combined heat and power (CHP) unit is used to heat office space it may be more useful to use heat meters as a means to establish energy use and to then allocate costs. CIBSE's *Building Energy Metering* (TM39) provides guidance for designers of new non-domestic buildings who wish to achieve the metering requirements of Part L of the Building Regulations, and provides good practice for the general application of energy metering in buildings.

When looking at production facilities, consideration has to be taken of the processes that are carried out and the manner in which the direct costs are allocated to the production stages. For example, if hot and cold water are used in the process, the volume of water used may be of greater interest than the amount of heat used. Similarly, the use of compressed air might lead to significant efficiency gains, but these will need to be compared with the cost of running the compressors against the volume of air used.

## Step 2: considering equipment that generates electricity

Where electricity is being generated on site by a CHP unit or a renewable system like solar photovoltaic (PV) or wind generation you will want to know how much electricity you are generating and how much you are using, and, if you have a feed-in-tariff (FiT) scheme, how much you are exporting.

If you have an installed generation capacity of 30 kW or less there is no requirement to have a specific export meter. Guidance on the legislation and the metering required for FiT schemes can be found on the Ofgem and the Department for Business, Energy & Industrial Strategy (BEIS) websites. (Please note the caveats as to the accuracy of this information.)

## Step 3: establishing the service to be measured

Once the energy challenge has been identified, the medium that you need to measure should become clear. In this example, the businesses that sub-let parts of the site could have their own gas, electricity and water meters. The commercial kitchen that uses gas could also have a gas meter. Heat meters could be used to measure the energy that is distributed from the CHP unit through the district heating system. In the main plant electricity meters could be installed to record how much power each department uses.

## Step 4: understanding the reasons for measuring the service

This step should be undertaken at the same time as steps 2 and 3.

A number of questions can be asked to understand exactly why the service is being measured.

### (a) Allocation of costs?
In this example, the business site comprises group companies and departments that are using utility services but can't measure how much they are using or what level of overhead they carry. Metering at a departmental level will provide that transparency and enable the site manager to allocate the costs to the right departments.

### (b) Monitoring usage
One of the most important uses for secondary metering is in the area of energy saving. The first stage in the process is to be able to measure the amount of electricity or gas that is being used and to keep a record of it so that a picture of usage can be created over time. Usage patterns can be monitored and trends can be identified. Without this stage it becomes more difficult to establish whether a change is having an impact.

In the example, the manufacturing company can install electricity meters in all the significant areas, such as manufacturing spaces, offices, and/or warehousing areas to monitor what is being used and when. There are a number of guides available that provide direction specifically in relation to energy savings, including the Carbon Trust *Monitoring and targeting – in depth management guide* (CTG008) and the CIBSE Energy Benchmarks (TM46).

### (c) Identifying savings

Once the utilities being used are monitored, opportunities to make savings can then be identified. In the example, it might be that the offices are being heated when they are not occupied or that storage facilities are being lit when nobody is working in them. Installing secondary metering enables the manager to break down the energy usage to finer elements. The more meters that are fitted the higher the level of accuracy.

### (d) Usage control

Installing secondary meters on a large site provides greater visibility over where utilities are being used and enables a greater level of control.

In the example, in order to encourage more awareness, the site manager can initiate a scheme for the departmental managers whereby they can have incentives for capping the amount of power used over a monthly cycle. This would not be possible unless their usage was measured independently.

### (e) Revenue creation

In the example, there are units on the site that are sub-let to other companies. The site manager, in capacity as a landlord, can resell gas and electricity through the use of secondary meters. Meters used for these purposes need to meet the same requirements as primary meters (see Section 3). The BPF Office ABI Landlord's Energy Statement Guidance and Specification Version 2.3 April 2007 provides detailed guidance and tools on establishing energy usage in sub-let commercial premises and allocating statements to users.

## 2.2 Legal requirements

The installation of meters is covered by a wide range of primary and secondary legislation. The specific requirements depend on the types of meters to be installed and their use. A full list of applicable legislation is provided in Annex A.

A critical area is fiscal metering, which is covered by specific legislation.

**All** gas and electricity meters used for billing **must** be approved, be it primary or secondary meters, either under national legislation or the MID.

Meters not used for billing (i.e. monitoring only) do not have to be approved, however, if this requirement should change and the end user is charged on their usage via the meter, it must be changed for an approved meter.

From 30th October 2016, all new meter designs that fall within the scope of the MID must be approved by the MID. UK approved meters may still continue to have modifications to the original approval and be placed on the market until the end of the 10-year derogation period, on the 30th October 2016.

National legislation meters already in service, or 'on the wall', may remain, as long as they continue to conform to the relevant legislation.

© The Institution of Engineering and Technology

The MID only covers meters *intended for residential, commercial and light industrial use*. These terms were never formally defined by the MID and it is up to each member state to define what these are in their respective implementing regulations.

### 2.2.1 Gas meters in the UK

For gas meters, the British national legislation and the MID cover everything up to and including a maximum flow rate of 1,600 m³/h under standard operating conditions, as laid out in the regulations.

Any gas meter with a maximum flow rate above this is not regulated and it is down to the end consumer to ensure that the meter is suitable and meets their requirements. These details are agreed within the gas supply contract and include measures surrounding accuracy and uncertainty.

### 2.2.2 Electricity meters in the UK

Electricity meters are different. The MID implementing regulations for the UK, the Measuring Instruments (Active Electrical Energy Meters) Regulations, states that meters do not fall within the scope of the MID and must still be nationally approved if:

**(a)** the maximum quantity of electricity supplied exceeds 100 kW/h; and
**(b)** the meter provides measurement on a half-hourly basis.

This is because such meters are likely to be heavy industrial-type meters.

Meters above 100 kW/h must also meet the additional requirements of the Balancing and Settlement Code, administered by ELEXON.

## 2.3 Who is going to be responsible for the meter?

The responsibility for metering could come under a range of roles such as energy manager, energy services consultant, safety, health and environment (SHE) director, environmental manager, etc.

What is more important is that it is clear who is responsible for each aspect of the installation, i.e. the design, installation, maintenance, data collection and reporting. Meter installations can represent a significant investment and the whole value of that investment will only be realised if all roles and responsibilities are clear.

## 2.4 How much data is required?

The volume and frequency of the data required will depend on what it is being used for. An energy dashboard is used to give near real-time display of how much energy is being used in a building or an organisation. To achieve this you need to be collecting meter data on a very frequent basis.

On the other hand, if a meter is being used to raise an invoice for electricity or gas used, you may be happy to collect the data monthly or quarterly. Normal reading rates range from half-hourly to daily. Of course, the higher the frequency of the reading the more data you will collect and the greater the data storage requirements.

## 2.5  Preparing the business case

Once it is clear what utilities are to be measured and how, it is time to put all the elements together and create the business case for installing the meters.

A good place to start is with a detailed plan of the scheme showing where all the meters are going to be installed, what they are measuring and why. The Carbon Trust *Metering technology overview* (CTV027) includes a useful schematic diagram of a metering plan. Once a plan is in place, costs can then be accurately calculated. At the same time a calculation of the expected benefits can be made and these can be supplemented by an income statement from any meters that are used to collect revenue. Finally, a benefits case and proposal can be drafted.

## 2.6  Are specific competencies required?

The core skills required to carry out the installation of meters are available in the mechanical and electrical services field, although Gas Safe registered engineers must be employed to carry out gas work.

For planning and implementation assistance a building services engineer or energy consultant may be engaged.

© The Institution of Engineering and Technology

# Electricity metering

## 3.1 Meter selection

### 3.1.1 Basics of electricity metering

There are two basic metering methods:

**(a)** whole current, whereby the meter measures the whole current taken by the load; and

**(b)** CT-metering, whereby a current transformer (CT) is used to measure the current.

There are a number of communications options that allow for the metering data to be collected remotely, over either a wired or a wireless connection.

Mechanical meters are now unlikely to be installed as new as electronic meters are available and have improved functionality at a much lower cost. Electronic meters typically have a life of 10 years or more. An asset management plan may be advisable.

Output options will also play an important part when selecting the meter and it is fair to say that, in some cases, it could be the priority.

Building management systems (BMS), serial communications, internet monitoring and wireless access features are available for data acquisition. Communications systems are covered in Section 7.

Secondary meters that are not used for billing are outside the scope of the fiscal metering standards but BS 8431 may be specified in the procurement documents. For energy management and similar applications, Class 1 is normally adequate.

### 3.1.2 Basics of CT-metering

The main purpose of a CT is to convert primary current of a high value into a lower secondary value that can be interpreted by a meter.

These secondary values are typically 5 A or 1 A and are inversely proportional to the primary value dependent on the ratio of the transformer, for example, 300 A/5 A, with maximum primary current being 300 A that would give a value of 5 A on the secondary side.

**Note:** It is generally accepted that when a CT is operating near its maximum rating, accuracy tends to tail off. You should select a CT to match the maximum operating current. It is always better for the equipment owners not to operate at lower current levels rather than close to the maximum limit in order to maintain a healthy electrical system.

The CT ratio must be selected such that it matches the maximum operating current of the system to be measured, for example, if the maximum operating system current is 600 A, select a 600 A/5 A CT if the meter users 5 A inputs.

© The Institution of Engineering and Technology

### 3.1.3 Burden

The CT and the cabling to be used in the metering system will place a load (burden) on the meter, which will have an effect on the overall accuracy of the meter. The person or company selecting and designing the metering system should take burden into account.

For example: the length and size of the cabling will place unnecessary load (VA) on the metering. Upsizing the cable can help where the meter is some distance away, however, using a 1 A secondary CT could also lessen the effect of line losses.

The effects of burden and mitigation measures should be considered when designing and selecting a metering system so that accuracy can be maintained at the level specified for the application that the metering system is being used for.

### 3.1.4 Terminology

Understanding the terms used when specifying a CT is important; for example, it is important not to get protection CTs mixed up with metering CTs. Protection CTs are built to withstand much higher currents and if you were to use a protection CT for measurement there is the risk that, should current levels ever rise, the metering equipment may not be able to thermally withstand such high currents. Annex C sets out a list of glossary terms.

### 3.1.5 Accuracy

For the majority of sub-metering applications, Class 1 accuracy is generally acceptable, i.e. the accuracy will be +/-1 % of the actual current being measured.

## 3.2 Meter technology types

### 3.2.1 Categories of electricity sub-meters

There are several ways of categorising types of electricity sub-meters.

▼ **Table 3.1**      Categorisation of electricity sub meter types

| Category | Type | Notes |
|---|---|---|
| Method of connection (see 3.2.2) | Directly connected | – |
| | Indirectly connected | Connected via a current transformer (CT) or current sensor (CS). |
| Method of operation (see 3.2.3) | Electromechanical | Considered obsolete. |
| | Static | Electronic operation. |

© The Institution of Engineering and Technology

| Category | Type | Notes |
|---|---|---|
| Mechanical format (see 3.2.4) | Panel mounted | DIN 96 × 96 format is popular. |
| | DIN rail mounted | Normally onto a top-hat rail to BS EN 50022. |
| | Wall mounted | Not common with sub meters but can be accomplished by fitting a DIN rail or panel mount meter into an enclosure. |
| | Modular meters | The format may vary, but these allow the user to add additional measurements after installation without major engineering work. |
| Measurement (see 3.2.5) | Energy (kWh) only OR with power (kW) | |
| | Multifunction | Also measures instantaneous parameters and reactive/apparent energies. |
| | Harmonic | Also measures basic power quality parameters such as total harmonic distortion (THD), harmonics and neutral current. |
| | Power quality | Includes measurements to IEC power quality standards. |
| Display (see 3.2.6) | Full display | This is normally limited to panel mounted, which will need to be viewed by the operator. |
| | Minimal display | One- or two-line display for configuration only. Read via communications links. |
| Communications (see 3.2.7) | No communications | Very unusual; typically a traditional domestic meter. |
| | Pulse output | – |
| | Digital communications | – |
| Meter accuracy (see 3.2.8) | Class index | – |
| Latest advances (see 3.2.9) | – | – |

**Note:** Abbreviations for CT/CS are used within this Guide for convenience. Not all may be generally recognised.

## 3.2.2 Method of connection

### (a) Direct connection
Direct connection is the favoured installation method for tariff meters with maximum currents up to ~100 A, although meters up to 160 A have been manufactured. The advantage in this case is low cost, particularly since the meter is installed next to the incomer cable.

For sub-metering, the disadvantage is that the power cables need to be run to the meter location, necessitating a shutdown to run large cables.

### (b) Connection via a CT
The CT reduces the nominal measured current to 5 A, which is then fed to the meter. Other nominal output currents are sometimes used. The typical CT is a toroidal steel core with copper wire wound around the core. The primary current cable passes through the central hole. These are also available as split-core. A disadvantage, particularly of split-core, is their size and weight.

CTs should be mounted close to the meter as resistive losses in the cable will affect accuracy.

CT outputs should **NEVER** be left open circuit as dangerously high voltages can be generated at the terminals.

### (c) Connection via a CS
A CS has a voltage (rather than current) output, typically 333.3 mV at nominal input current. In fact, it is an ordinary CT terminated by a resistor. Since it does not have to drive a high current down a cable, the CS can be optimised for size and weight. Also, there are no risks associated with leaving the output open circuit and longer cables can be used. These are normally available as split-core, although ring types are manufactured. A major advantage over traditional CTs is their small size and light weight.

### (d) Connection via a Rogowski CS (Flexi)
A Rogowski is an air-cored current sensor. The benefit is that installation is simple – just wrap it around the cables. They do, however, require an electronic interface.

## 3.2.3 Method of operation

### (a) Electromechanical (Ferraris)
This is the traditional type of meter with a rotating disc. These are effectively obsolete but many remain in operation as they have a very long service life.

### (b) Static/electronic
All modern meters are static, using either a custom IC or custom firmware in a general purpose microcontroller. Current production is effectively all digital but very high accuracy and stability could be achieved using analogue techniques.

## 3.2.4 Mechanical format

### (a) Panel mounted
A very popular format, commonly in a DIN 96 × 96 case size although there are some meters with smaller enclosures. A key safety point is that the meter is mounted in an enclosure and there must be no access to the interior when power is applied.

© The Institution of Engineering and Technology

## (b) DIN rail mounted

Also known as TS35, this is mounted on a top hat type rail in accordance with BS EN 50022. The mounting is simple and conveniently placed at the rear of an enclosure. A display is provided for commissioning purposes but would not normally be visible from the exterior. Again, there must be no access to the interior when power is applied.

## (c) Wall mounted

Not common with sub-meters but can be accomplished by fitting a DIN rail or panel mount meter into an enclosure.

## (d) Modular meters

A multi-channel meter where the number of measurement channels can be increased (or reduced) after installation without any engineering work. The extra channels can be clipped on or plugged in at any time (although all power must be disconnected). The benefits are a single voltage and communications connection for up to 60 measurement channels.

▼ **Figure 3.1** Typical panel mounted applications (images courtesy of Andy Baker, Association of Meter Operators (AMO))

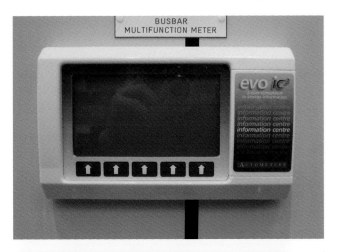

### 3.2.5 Measurement

#### (a) Energy (kWh) only, OR with power (kW)
This is a basic kWh meter, with a display that is typically LCD although there are some low-cost imports with a mechanical register. Addition of kW display greatly simplifies commissioning.

#### (b) Multifunction meter
In addition to kWh, multifunction meters can measure instantaneous parameters, reactive/apparent energies and other characteristics.

#### (c) Harmonic meter
A multifunction meter that also measures basic power quality parameters, such as THD, individual harmonics and neutral current. **The level of harmonics measured can vary according to manufacturer and the product specified.**

#### (d) D Power quality meter
Harmonic meter that also measures voltage swells, dips and interruptions, flicker to IEC 61000-4-15, etc. Not normally required unless there is an expectation of supply problems or high harmonic generation.

### 3.2.6 Display

Although meters without a display have been manufactured, these are not recommended. Installation and commissioning without a display can be lengthy and very frustrating.

#### (a) Full display
This is normally limited to panel mounted meters, which will need to be viewed by the operator. With multifunction and harmonic meters, there may be over 100 different display pages. In some cases, not all parameters will be available for display.

#### (b) Minimal display
This meter has a one- or two-line display and is generally used for configuration and commissioning. It is read via communications links.

### 3.2.7 Communications

#### (a) No communication
Very unusual and of little real value unless it is mounted where it can be clearly seen.

#### (b) Pulse output
Legacy system for remote reading of meters. The voltage-free pulses would be counted at an outstation. To ensure consistency, meter reading and pulse counts would need to be reconciled at regular intervals.

#### (c) Digital communications
Meter is connected to a SCADA, BMS or other system. A common UK protocol is Modbus RTU®, using RS485 over twisted pair cables. For reliable operation, wiring should be from meter to meter with no T-legs. Modbus TCP® is the Ethernet version.

Alternative protocols include M-Bus (commonly used on heat meters) and BACNet – used in BMS.

Meters are increasingly being directly connected either to an intranet or via the internet. Web pages are provided for configuration and the reading of data can be logged and

© The Institution of Engineering and Technology

sent to a remote server at regular intervals. Logging intervals down to 1 s are feasible, as is a 100 ms data update.

### 3.2.8 Meter accuracy

Meter accuracy is given as a class index, typically Class 1 for sub-meters. This means that the accuracy is ±1 % of reading under standard conditions and over a measurement range from 5 % to 120 % of the nominal rated current. Larger errors are permitted down to 2 % and where the power factor is not unity.

The reference standards for meter accuracy are the BS EN 50470 series (European standard for electricity meter accuracy), IEC 62052 and 62053 series of standards or BS 8431:2010. Other standards do not apply or are obsolete.

Current sensors will add to these errors. Under worst case conditions with 5 % current and a power factor of 0.5 inductive, an error of 11.9 % of reading would be allowable. In practice, the actual error would be very much less. Also, these errors are consistent; since for energy management we are comparing pre- and post-ante measurement, the errors for both sets of readings would be comparable.

### 3.2.9 Latest advances

Advances in metering come in two forms: higher specifications and reduced total installed cost. In most cases, the cost of the meter itself is only a small part of the total cost – particularly where the meters are being retrofitted. One of the biggest changes over the past 10 years has been the introduction of the miniature current sensor; not only are they very quick to install but they are much lighter and require less mechanical support.

Recently, the availability of 3Φ block current sensors that simply plug in to the meter has provided cost savings for new installations. Following on from this is the patented smart CT that communicates its specification to the meter at power-up.

▼ **Figure 3.2** Example modern panel mount meter cube 950 V with 3 × 3Φ measurement channels (images courtesy of ND Metering Ltd)

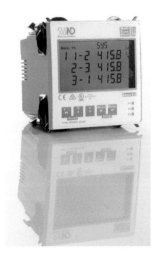

**▼ Figure 3.3**   Examples of DIN rail meter mounted in a standard MCB enclosure for wall mounting

*Courtesy of ND Metering Ltd*          *Courtesy of Secure Meters*          *Courtesy of Iskraemeco*

**▼ Figure 3.4**   Example DIN rail mounted meter (image courtesy of ND Metering Ltd)

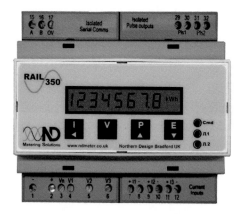

**▼ Figure 3.5**   Example modular meter (with 8 × 3Φ channels) plus remote display (image courtesy of ND Metering Ltd)

© The Institution of Engineering and Technology

## 3.3 Current sensor and current transformer technology

### 3.3.1 Current sensors

Most sub-meters do not directly measure current; they use external current sensors (CSs). These can be the traditional CT with an AC output, but more commonly with modern meters CSs with an AC voltage output. Each type has its advantages and disadvantages depending on the application.

### 3.3.2 Current sensors: errors

There are two types of errors associated with all current sensing devices:

**(a)** ratio error, which is the ratio-metric error between the primary and secondary currents; and
**(b)** phase error, which is the phase difference between the primary and secondary signals.

These errors vary in accordance with changes in the input current; the amount of variation permitted depends on the accuracy of the sensor.

### 3.3.3 Current transformers

These are passive electromagnetic devices, commonly toroidal in shape. The primary current passes through the middle of the toroid inducing a varying magnetic field in a steel core. This changing field induces a current in the secondary winding, normally of multiple turns. The current is reduced by the ratio between the turns.

The secondary circuit of a CT should never be left open circuit where there is any possibility of the primary circuit being energised. If the primary circuit is energised, a high voltage will develop across the secondary circuit. Depending on the current and the design, these voltages may be dangerous.

When selecting CTs, the impedance of the secondary circuit needs to be taken into account. CTs are designed for a defined accuracy at a maximum secondary load (for example, Class 1 at 5 VA). If this load is exceeded, the accuracy will be affected. For a 5 A CT, the applied burden of a 5 m run of 2.5 mm$^2$ cable is 2 VA; to this should be added about 0.5 VA for each set of terminals plus the meter burden. For a 1 A CT, these burdens are reduced by a factor of 25 ($V = I^2R$).

For metering applications, an accuracy of Class 1 or better, in accordance with IEC 61869-2, is required. Note that the accuracy of a Class 1 CT is not specified below 5% of nominal current and that, under nominal conditions, the worst case error of Class 1 CTs with a Class 1 meter is 12.4 % of reading.

CTs are available in three different formats.

| Format | Description |
|---|---|
| Toroidal | The primary current passes through the middle of the toroid, effectively forming a single turn.<br><br>Typically used on currents from ~100 A. |
| Wound primary | Multiple primary turns are wound round the core, fully insulated from the core and the secondary windings.<br><br>These are used for currents below 100 A. |
| Split core | These are toroidal transformers with a cut steel core. This simplifies installation on retrofits but size and weight may cause difficulties |

### 3.3.4 Current sensors

These can be either passive or active devices, normally with a voltage output; some may require a special interface. A common output is 333.3 mV at nominal FS current, although other outputs may be used.

CSs are available using various different technologies.

▼ **Table 3.3**     Current sensor technologies

| Type | Description |
|---|---|
| Passive | A CT with an integral burden resistor. These are manufactured either as split core or as a ring.<br><br>As effective as a complete system, the design can be optimised for size and accuracy.<br><br>For example, a 100 A/5 A 2 VA Class 1 split core CT weighs between one and two kg while a CS with similar specifications weighs approximately 100 g. |
| Rogowski | An air cored device, frequently wound onto a flexible circular base. It can thus be installed by winding it around the primary cable(s). The output voltage is very low and phase-shifted by 90 degrees in relation to the measured current. The output thus needs to be integrated and amplified. |
| High voltage | Not typically found in a sub-metering application except for larger operators. |

▼ **Figure 3.6**     Examples of current sensors/current transformers (images courtesy of ND Metering Ltd)

(a) Traditional tape wound CT

(b) 100 A/5 A traditional CT with a miniature 100 A:333.3 mV CS

© The Institution of Engineering and Technology

(c) 3Φ block CS with plug-in RJ connector

(d) 3-phase set of Rogowski sensors

(e) Clamp-on CTs and CSs for portable applications.

# 3.4 Installation

### 3.4.1 Site survey

A thorough survey and planning stage is essential before any installation work begins. This should be carried out by a person with appropriate skills, electrical knowledge and experience.

Site surveys are fundamental for a successful meter installation. These can take days to undertake and, if not done thoroughly, can cause problems for the installer.

An accurate site survey will pay dividends by:

**(a)** keeping the shut-down time to a minimum;
**(b)** keeping the impact on customers' down time to a minimum;
**(c)** minimising service disruptions; and
**(d)** avoiding unforeseen circumstances.

### 3.4.2 Existing electrical infrastructure and equipment

It is important to understand the architecture of the installation and how the electrical infrastructure is laid out across the site. A decision about where to install the meters cannot be made without carrying out this survey.

Knowledge of the electrical equipment used on site will help best define the use of metering. Access to manufacturer's information about the electrical equipment present would be advantageous in this respect and will help to cut down the time on site and reduce the risk of intrusive surveys, which may require some of the installation to be temporarily shut down. More importantly, it will reduce the exposure of personnel to safety risks when they undertake this work.

In some cases there may be meters already installed. Careful study of electrical drawings and manufacturer's data can help identify whether equipment may already be fitted with meters.

From the results of the site survey one can start to piece together the form of the metering that needs to be installed.

### 3.4.3 Installation of secondary meters/sub-meters

Consideration should be given to the environment in which the meter and associated equipment are to be installed. Secondary meters are normally manufactured for internal use only. For external applications a suitable enclosure should be used.

All electrical wiring to secondary meters must comply with the latest edition of BS 7671. Meters must be installed by suitably qualified and experienced persons. Due consideration should be given to where the meters are placed because someone may have to visit the site in the future to read, inspect or maintain them. In particular, if possible, avoid installing meters at heights that would necessitate the need for access equipment to maintain them.

Cable runs, cutting of panels, drilling and close proximity to live electrical services play a significant role when installing meters.

### 3.4.4 Site surveys for current transformer installation

Site surveys for CT installation are important so that issues can be understood and overcome during the installation. Such issues may include:

**(a)** space requirements;
**(b)** access for future maintenance;
**(c)** where to locate the voltage connections and fuses;
**(d)** where to locate the current pickups; and
**(e)** assessing what is already installed.

Safety is always important and using a competent person to carry out site surveys will help to identify whether solid bars or cables are fitted, which will determine the type of CTs to be used. In some cases bespoke CTs may have to be installed.

The survey should enable the installer to determine how long it will take to complete the installation and whether the client will need to provide additional services, such as telephone or data lines. It will also enable the installer to advise the client on whether they should consider out-of-hours working or shutdowns.

## 3.5 Commissioning

### 3.5.1 Role of commissioning

Commissioning is fundamental for a successful meter installation. Experience has shown that, where something has gone wrong, it is usually down to either the installation or the commissioning, for example:

**(a)** measuring equipment has been fitted the wrong way around P1/P2;
**(b)** wiring has been crossed S1/S2;
**(c)** the meter has not been correctly programmed – 5 A secondaries instead of 1 A;
**(d)** interference with other services has distorted the results; and
**(e)** cables connections are poor.

© The Institution of Engineering and Technology

**▼ Figure 3.7** Example of a metering schematic

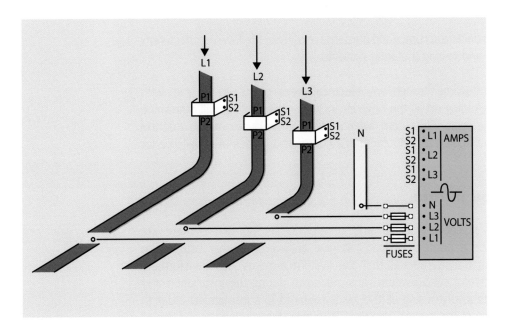

## 3.5.2 Common errors and corrective actions

### (a) Wiring
To measure electric usage you need to measure both voltage and current but also make sure that the wiring is correct when measuring each. This is particularly important when measuring 3-phase electric circuits. Care should be taken, as voltage connections and current connections can be bundled together. If not correctly segregated and arranged, a simple phase-shift by 120 degrees would cause errors to occur in the metering system, which can affect operation (billing/monitoring) and understanding of the relationships between waveforms if the metering is being used to measure power quality.

### (b) Direction/polarity
Fitting the CT in the direction of the primary current flow is the first step (P1 and P2) but also wiring the secondary connections – see 'crossed secondary CT wires' below.

### (c) Fusing
Fuses should be as sized and selected by the designer and installed as appropriate to the application that is being metered.

### (d) Cable sizing
*See also* Burden.

Measure primary and secondary outputs. These outputs should have been tested and documented during the commissioning process.

### (e) Wrong phasing
Where voltage from one phase L1 is being measured with current from another phase C2, isolate all the phases and work on one phase at a time to correct the wiring.

### (f) Crossed secondary CT wires
Crossed secondary CT wires will either give an incomplete reading or no reading at all. All CTs should have S1 and S2 shown on the output side. If crossed, these will cause current to flow in the wrong direction in the metering and hence be given a wrong polarity – to fix, isolate and work on one phase at a time, correcting any crossed wiring.

© The Institution of Engineering and Technology

### (g) Incorrect ratio setups

Use a handheld current clamp and compare against the meter reading.

Manufacturers will always declare accuracy and burden levels. Always follow manufacturer's instructions when fitting and testing metering systems.

Remember that current flowing through any electrical piece of kit will be met with resistance (impedance). Adding up all the devices, including the cabling, will determine the overall burden of the metering system, which needs to be overcome so that the required accuracy can be achieved. Manufacturer's charts and guides are available for reference.

## 3.5.3 Summation setups

Summation setups are available, cheaper and quicker to set up where there are applications that combine a number of loads to give an overall energy usage. The advantage of summation transformers is that the system requires less transformers and cabling. However, careful consideration must be taken during the installation and commissioning. The tests involve using all the loads required to summate, in order to prove the overall accuracy of the system plus the CT primary flows are balanced, i.e. all of the loads should be similarly rated in order to give reasonable results.

In typical applications, summation CTs are usually found on main distribution boards where current is being measured from two or more incomers and the one transformer measures the net result of all these incomers.

More information is provided in Annex D.

© The Institution of Engineering and Technology

### 3.5.4 Harmonics and metering

**A brief explanation**

There are many publications and online resources available about harmonics and ways of dealing with them. This Guide will cover the basics and typical methods of dealing with harmonic distortion.

**Causes of harmonics**

The basic causes of harmonic distortion are non-linear loads that alter the normal sine wave (voltage or current) of the AC electrical supply. Typical electrical loads that cause harmonic distortion include switch mode power supplies, variable frequency drive devices and other modern electronic equipment with built-in switching components that manipulate the electrical supply.

**Effects of harmonic distortion**

Harmonic distortion can cause:

**(a)**    increases in neutral currents;

**(b)**    opposition to normal flow of currents; and

**(c)**    distortion of voltage.

These effects can lead to issues such as overheating and stresses, which, if serious, could damage equipment, reduce its lifetime or potentially cause a fire.

Problems that harmonic distortion can cause include:

**(a)**    disturbance to IT infrastructures, including power disturbances;

**(b)**    reduced efficiency of electrical equipment such as motors;

**(c)**    capacitor burnout that impacts power factor correction; and

**(d)**    equipment malfunctions.

It is always important to undertake site surveys and, if possible, incorporate power quality measurements in the surveys. This will give the benefit of having upfront information, which will help in deciding the metering that should be employed.

**Solving harmonics**

Use inline reactors in series with the supply and fit, usually, prior to non-linear load. Such reactors look very similar to a transformer but instead of being wired in parallel are wired in series. Transformers can be used and give an added advantage of providing isolation, however, these come at greater cost.

Reactors are cheaper and provide a form of filtering in terms of limiting the disturbances or current.

## 3.6  Operation and maintenance

Generally meters do not require maintenance, have no parts that are accessible to the user and should be replaced on failure. Some meters may be equipped with batteries that are designed to protect data storage against power loss. Occasionally, these batteries may need replacing. More often than not the complete meter would be changed in these circumstances.

## 3.7  Decommissioning and disposal

Meters contain hazardous materials and must be disposed of through an approved process to comply with the requirements of the WEEE directive. Licensed disposal companies must be used.

© The Institution of Engineering and Technology

# Gas metering

Secondary or sub-meters are commonly used in gas installations. In the past, on large installations, it was common for a second primary meter to be installed downstream of the main fiscal meter to enable a specific load to be measured and invoiced separately, for example, where there was a special contract load. These arrangements were referred to as 'sub-deduct meters' and they are no longer allowed.

Now, all primary meters have to have their own unique supply. In order to avoid confusion, secondary meters that are fitted downstream of the main primary meter but are not used for billing are called 'check meters'. Guidance on the installation of check meters can be found in BSI 6400 Part-1:2016 *Specification for installation, exchange, relocation and removal of gas meters with a maximum capacity not exceeding 6 m3/h – Part 1: Low pressure (2nd family gases)* and in the IGEM standard IGE/GM/8 *Non-domestic meter installations. Flow rate exceeding 6m3h-1 and inlet pressures not exceeding 38 bar.*

## 4.1  Meter selection

### 4.1.1 Pressure range

A meter should not be used outside of its designed and badged pressure range. Table 4.1 provides guidance on the pressure ranges available.

### 4.1.2 Flow range

The flow being measured by a meter should be, as far as possible, within its flow range ($Q_{min}$ to $Q_{max}$) as specified by the manufacturer. If a high level of confidence with the accuracy of the reading is required, the maximum flow rate should not exceed the meter badged maximum flow.

No meter should be over-sized in order to accommodate high flow rates that may occur very infrequently. However, consideration should be given to possible short-term over-speed effects. Under such circumstances, overload flow rates should be limited to 120 % $Q_{max}$. Pressure absorption limitations may result in some over-sizing of the meter being necessary but over-sizing a meter could significantly reduce the effective turndown of the meter. This may result in an under-registration of the gas supplied. An example of the type of situation for which the meter should not be over-sized would be an excessive flow that might occur during the commissioning of a certain plant, such as a gas compressor. In some cases, it may be necessary to restrict the way in which the plant operates to avoid over-speeding the meter.

## 4.1.3 Factors affecting gas meter selection

▼ **Table 4.1**    Factors affecting gas meter selection accuracy and performance

| Factor | Diaphragm | Rotary displacement | Turbine | Multipath ultrasonic |
|---|---|---|---|---|
| Range of $Q_{max}$ | 6 to 160 m$^3$ h$^{-1}$ <br><br> 212 to 5650 ft$^3$ h$^{-1}$. | 25 to 2,885 m$^3$ h$^{-1}$ <br><br> 800 to 102,000 ft$^3$ h$^{-1}$. | 65 to 25000 m$^3$ h$^{-1}$ <br><br> 2275 to 882,875 ft$^3$ h$^{-1}$. | 65 to 85000 m$^3$ h$^{-1}$ <br><br> 28,000 to 3001,900 ft$^3$ h$^{-1}$ |
| Typical pressure range | 0 to 75 bar. <br><br> Others available as special case | 0 to 10 bar. <br><br> Special meters available up to 38 bar. | 0 to 38 bar. | 0 to 38 bar. |
| Typical range ability (and accuracy) | Badged: 50:1 (±2 % to ± 3 %) <br><br> Usable: > 150:1 (±2 % to ± 3 %) <br><br> Dynamic: > 1000:1. | Badged: 20:1 to 50:1 <br><br> (±1 % to ±2 %) <br><br> Usable: > 50:1 <br><br> (±1 % to ±2 %) <br><br> Dynamic: 500:1. | Badged: 10:1 to 30:1 <br><br> (1 % to 2 %) <br><br> Usable: > 10:1 <br><br> (±1 % to ±2 %) <br><br> Dynamic: 75:1. | Certified over 40:1 to 125:1 (±1 %) <br><br> Usable: > 40:1 (±1 %) <br><br> Dynamic: Not applicable. |
| Effect of gas density | Unaffected in design range within manufacturer's specification. | Insignificant. | Minimum flow is lowered with increased density, increasing the usable and dynamic range ability. | Meter accuracy does not deviate over the specified working range of transducers. Certain types of transducer will not operate at low densities, depending on meter size, line density and gas composition. |
| Effect of solids | Normally unaffected but coarse filter is recommended at higher pressures. | Meter may stop rotation. Filter required. | Blades may be damaged and freedom of rotation may be affected. Coarse filter required. | Normally unaffected, but contamination of the transducers can affect meter performance. |

© The Institution of Engineering and Technology

| Factor | Diaphragm | Rotary displacement | Turbine | Multipath ultrasonic |
|---|---|---|---|---|
| Effect of gas-borne liquids, e.g. water, oil, grease etc. | Corrosion possible. | Corrosion possible. | Corrosion possible. | Liquids settling in the bottom of the meter or grease on the internal walls reduce the cross-sectional area and cause the meter to over-read. |
| | Freezing may result in damage. | Oil may be displaced from gears. Freezing may stop the meter. | Freezing may result in damage. Lubricant dilution and rotor imbalance possible. | Freezing may cause a temporary increase in uncertainty. |
| | Materials of construction may be affected. | Materials of construction may be affected. | Materials of construction may be affected. | Materials of construction may be affected. |
| | Over-registration possible. | Under-registration possible. | Inaccuracy possible. | |
| Pressure variations | Excessive differential pressure variations will cause damage. | Rapid change of differential pressure may cause damage. | Rapid pressure changes may cause damage or registration errors. | Normally unaffected. |
| | | | Particular problems when meters are installed inter-stage at higher pressures. | |
| Acoustic noise | Unaffected. | Unaffected. | Unaffected. | Can be affected by acoustic noise. |
| | | | | Precautions need to be taken with the location of the meter and its proximity to noise sources such as control valves, pressure regulators and partially open line valves. |

The flow range specified is related to the actual flow through a meter at the prevailing pressure and temperature. Consequently, metering pressure has to be taken into account when selecting a meter. The flow range also affects the accuracy of the meter and needs to be taken into account. When a meter is installed upstream of the pressure regulator, any variation in inlet pressure to the meter will have a significant impact on the turndown of the selected meter. The meter will have to cope with the maximum flow at the minimum pressure and the minimum flow at the maximum pressure. A 2:1 variation in pressure will result in a halving the usable turndown of the meter.

## 4.2 Effects of load

### 4.2.1 Diaphragm meters

In general, diaphragm meters are unlikely to be affected by downstream load changes. However, diaphragm meters should not be installed where they are subject to vibration and continuous pressure pulsations. Both of these effects can be caused by boosters or compressors and may cause the valve mechanism to 'bounce', thus giving rise to substantial under-registration.

### 4.2.2 Rotary displacement meters

A rotary displacement (RD) meter can cause operating problems in the following situations and should be used with caution:

**(a)** where large-step load changes are anticipated;
**(b)** where a booster (or compressor) is installed; or
**(c)** where small burners, for example, permanent pilots, are also supplied.

In extreme cases, an increase in sudden load may cause a temporary low pressure at the meter outlet, resulting in pilot outage or burner system lock-out. A sudden load decrease may cause the meter to over-run, resulting in a temporary over-pressure condition with consequences similar to the low pressure condition. These problems are due to meter inertia and may be minimised by installing a large reservoir of gas between the meter installation and the appliances or, in some cases, the installation of a suitable non-return valve at the outlet of the meter installation.

Where possible, the use of an RD meter for such loads should be avoided; instead, a diaphragm meter or, in some cases, a turbine meter should be used as an alternative. Where the use of an alternative meter type is not possible, a dynamic analysis model should be performed to determine whether the expected transient pressure changes are within acceptable limits.

RD meters generate small pressure pulsations. If the meter is subjected to pressure pulsations of similar frequency (or simple harmonics thereof) to the inherent meter pulsation, the system may resonate. Substantial metering errors may arise if resonance occurs and, if the condition persists, the meter may be damaged.

### 4.2.3 Turbine meters

A turbine meter should not be used to measure flows that are rapidly pulsating, nor should one be used where the total metered gas flow is on/off, unless the 'on' time is greater than 30 minutes. The necessary 'on' time can be reduced to two minutes if a continuous base load flows through the meter equal to at least 10 % of the maximum metered flow. This is because the turbine wheel continues to rotate for some time after the flow through the meter has ceased, which can result in substantial errors.

With certain turbine meters, the above procedures may be relaxed and shorter 'on' times would be acceptable, provided the meter manufacturer gives an assurance that metering errors due to non-steady flow will not exceed 1 %. Pulsating flows may be caused by cycling loads having on/off controls or even by some high/low controls. In addition, reciprocating gas compressors and engines will give rise to flow pulsations, which would affect the metering accuracy and, in extreme cases, can reduce operational life or cause damage.

© The Institution of Engineering and Technology

Where it is not possible to use an alternative type of meter, steps should be taken to ensure that flow oscillations at the meter are reduced to within the limits specified in the Institution of Gas Engineers and Managers' (IGEM) standard for the application of compressors on natural gas fuel excess noise is not generated at systems IGEM/UP/6. Where flow pulsations are suspected, a dynamic analysis should be performed to verify that the pulsations have been reduced to an acceptable level.

## 4.2.4 Ultrasonic meters

In general, ultrasonic meters (USMs) are unlikely to be affected by downstream loads. However, consideration should be given to the particular 'measurement and computational' strategy employed, as certain combinations of strategy and load characteristics may result in metering errors.

Pressure regulators and control valves generate noise in the frequency range that can affect the operation of USMs. Care should be taken when designing systems to ensure that excess noise is not generated at high frequencies or, where this is unavoidable, that the installation is designed to accommodate noise reduction equipment.

## 4.2.5 Ancillary equipment

Meters can be supplied, or be capable of being fitted, with a low frequency (LF) transmitter system and/or telemetry. Some meters can be supplied with encoding indexes that provide actual meter readings and include communications technology.

Any electrical connection to a meter should be made in accordance with the IGEM standard for electrical connections for metering equipment (IGEM/GM/7A).

▼ **Figure 4.1**   Typical meters for Check Meter Applications

*Courtesy of Elster Metering Limited*

*Courtesy of FMG*

*Courtesy of Elster Metering Limited*

*Courtesy of FMG*

© The Institution of Engineering and Technology

*Courtesy of Elster Metering Limited*     *Courtesy of Secure Meters*

### 4.2.6 Red secondary meters

In the past, there was an understanding in the gas industry that secondary or sub-meters were painted red to help meter readers distinguish them from primary meters. These were often installed in tenanted properties and were used to sell gas to tenants.

▼ **Figure 4.2**   Illustration of a red secondary meter

# 4.3  Installation

## 4.3.1 General

The primary objectives are to ensure that, as far as possible:

**(a)** pressure is managed so that a suitable pressure is supplied to appliances.

**(b)** measurement is continuously accurate, with both random and systematic metering errors reduced where this can be justified economically. The overall accuracy of a meter installation will depend upon the type of meter installed, the method of any volume conversion being used and the configuration of the installation.

**(c)** the operation is reliable and the required level of security of the gas supply is achieved.

**(d)** persons and plant are safe.

© The Institution of Engineering and Technology

Any meter and its pressure regulator installation, if required, should be designed not in isolation but as a single unit. Both should be sized for the same load and consideration should be given to the effect that each may have on the other. The selected meter should be capable of matching the characteristics of the load with the required degree of accuracy. The pressure regulator installation should normally be located upstream of the meter. The standard of filtration provided should give adequate protection for the PRI/regulator, the meter and any associated equipment.

## 4.3.2 Accuracy

There is no specific regulation governing the accuracy of check meters, however, as a guide the accuracy bands quoted in the MID could be used to specify a meter.

## 4.3.3 How to calculate the energy used and volume conversion

Gas meters measure the volume of gas that flows in cubic meters or cubic feet. However, to get an accurate measurement of the energy used, this will need to be converted to Kilowatt hours (kWh). This can be done using the following steps:

### Step 1

Subtract your previous reading from your current reading to give the number of units used.

### Step 2

The conversion factor from cubic feet to cubic metres is 0.0283, therefore:

1 cubic foot = 0.0283 cubic metres

100 cubic feet = 2.83 cubic metres

For imperial meters the reading (in hundreds of cubic feet) is therefore multiplied by 2.83 to convert to cubic metres (if the reading to the supplier was provided in cubic feet then this should be multiplied by 0.0283). This step is not required for metric meters as they are read directly in cubic metres.

### Step 3

This figure is then multiplied by the calorific value of the gas, which is a measure of the available heat energy. Calorific values vary and the figure quoted on your bill (for example, 39.5 megajoules per cubic metre ($MJ/m^3$)) will be an average of the gas supplied to the property.

### Step 4

The figure is then multiplied by 1.02264 to correct the volume of gas to account for temperature and pressure (as gas expands and contracts).

### Step 5

Finally, the figure is converted to kWh by dividing by 3.6.

For larger installations the meter reading must be adjusted to account for the effects of temperature, pressure and, on very large installations, compression. This is done by installing an electronic conversion system with the meter installation. Guidance on specifying and installing converters can be found in IGEM standard IGEM/GM/5 *Electronic gas volume conversion systems*.

▼ **Figure 4.3** Typical gas volume conversion system

*Courtesy of Elster Metering Limited*

## 4.3.4 Pressure

The installation must be designed to maintain a suitable pressure for downstream appliances to ensure their safe operation under all foreseeable conditions. Often, check meter installations will not require pressure regulation as the pressure will be controlled at the primary meter. On large installations with diverse loads it is not uncommon for a low pressure supply to be taken from a higher pressure feed, for example, the installation of a catering unit in a factory.

Reliable information relating to the nature of the load should be obtained and should include:

**(a)** the estimated maximum flow rate (which is not necessarily a summation of the total connected load).

**(b)** the minimum flow rate anticipated (a realistic assessment and not a zero flow rate).

**(c)** the number and type of each unit of plant and, where available, the anticipated load pattern for each.

**(d)** the best possible estimate of the anticipated growth of load over the next 12 months.

**(e)** where necessary, the peak flow that is likely to occur for a short period, infrequently, possibly in excess of $Q_{max}$.

**(f)** establishment of whether pressure control is required and, if so, what pressure the PRI/regulator(s) is/are required to control.

**(g)** STP (for the downstream system and all connected plants).

**(h)** Design Pressure (DPc), if higher than Maximum Operating Pressure (MOPc).

**(i)** Maximum Incidental Pressure (MIPc) (to which the meter installation could be subjected by the gas consumer's plant and equipment).

**(j)** MOPc (for which the downstream system has been designed to ensure that all connected plants will operate with an inlet pressure not exceeding $P_{max}$).

© The Institution of Engineering and Technology

**(k)** Operating Pressure (OPc).

**(l)** Lowest Operating Pressure (LOPc) (for which the downstream system has been designed to ensure that all connected plants will operate with an inlet pressure exceeding $P_{min}$).

**(m)** Design minimum pressure (DmPc) (for which the downstream system has been designed to ensure that all connected plants will operate safely and which may have to be obtained from other sources). Details of any special features of gas plants that may affect the nature of the load, for example, fast-fluctuating loads, snap-acting control valves creating rapid on/off load conditions, engines, turbines, gas boosters and compressors, should be obtained from appropriate sources.

**Note:** All pressure settings are on the consumer side, which means that they are downstream of the primary meter installation so that they would be identified by a subscript 'c', i.e. MOPc.

## 4.3.5 Pressure loss

One of the biggest problems with check meter installations is pressure loss. If the meter is sized correctly for the load, the act of installing the meter in line with the pipe system will render the pipe system out of standard for total pressure loss from the primary meter outlet to the appliance. A common reason for poor pressure at the appliance is badly designed installation pipework, which would only be exacerbated by the installation of a meter. One way to address this is to over-size the feed pipework and/or meter installation to minimize the impact. However, care should be taken to ensure that the meter will still measure across the range of flow required. The most important aspect is that the installation of a secondary meter should not adversely affect the performance of appliances in the most adverse conditions.

For a meter operating at an inlet pressure of 21 mbar, at the required $Q_{max}$, the pressure loss across the meter should not exceed 1.25 mbar on gas, unless it is known that a higher pressure loss is acceptable and safety and operational requirements are not compromised.

## 4.3.6 Gas consumer-imposed constraints and special requirements

The introduction of smart meters in residential installations imposes obligations on energy suppliers in respect of meeting the requirements of the Gas Safety (Installation & Use) Regulations. In accordance with the regulations, it is an offence to install a secondary meter downstream of either a pre-payment meter or a smart meter in pre-payment mode. If a secondary meter is installed downstream of a smart meter in credit mode the gas supplier should be advised so that they can amend their records.

It is vital that any other constraints imposed by the consumer are established, particularly if it may affect safety. In addition there may be special requirements for measurement, for example, to account for the EU Emissions Trading Scheme. This should be taken into account when planning the validation of the performance of the meter installation.

Consideration should be given, at the design stage, to the following:

**(a)** the intended location of the meter;

**(b)** appropriate means of access;

**(c)** the method of handling and construction;

**(d)** maintenance;

**(e)** the protection of the downstream system;

**(f)** velocity;

**(g)** noise; and

**(h)** heating.

## 4.4 Provision and arrangement of components

### 4.4.1 Gas meter installation standards

For meters with a $Q_{max}$ 6 m³/h, BS 6400 provides direction for the standard of installations of check meters. For other sizes, IGEM standards should be referred to. The standards for primary meter installation for guidance can be followed, as outlined in Table 4.2.

The gas flow through the meter installation should be from left to right (when viewing the index of the meter) and any meter should be positioned in such a way that the index can be read conveniently, without the use of mirrors, etc.

▼ **Figure 4.4** Typical secondary/check meter installations using turbine meters (images courtesy of Cameron Steel, BK Design Associates UK Ltd)

© The Institution of Engineering and Technology

▼ **Table 4.2**    Gas meter installation standards

| Installation details | Description |
|---|---|
| $Q_{max} \leq 6$ m³/h<br>MOP $\leq 75$ mbar<br>Standard installation | Domestic meter installations supplied at low pressure and installed in accordance with BS 6400-1. |
| $Q_{max} \leq 6$ m³/h<br>75 mbar $<$ MOP $\leq 2$ bar<br>Standard installation | Domestic meter installations supplied at medium pressure and installed in accordance with BS 6400-2. |
| 6 m³/h $< Q_{max} \leq 40$ m³/h<br>MOP $\leq 75$ mbar<br>Standard installation | Domestic meter installations up to U40 supplied at low pressure and installed in accordance with IGEM/GM/6. |
| 6 m³/h $< Q_{max} \leq 1076$ m³/h<br>MOP $\leq 75$ mbar<br>Standard installation | Industrial and commercial meter installations supplied at low pressure and installed in accordance with IGEM/GM/6. |
| $Q_{max} > 6$ m³/h<br>75 mbar $<$ MOP $\leq 38$ bar | Industrial and commercial meter installations supplied at pressures up to and including 38 bar, with a metering pressure of 21 mbar and installed in accordance with IGE/GM/8. |
| $Q_{max} > 6$ m³/h<br>75 mbar $<$ MOP $\leq 100$ bar<br>Non-standard installation | Industrial and commercial meter installations supplied at pressures between 16 bar and 38 bar and installed in accordance with IGE/GM/8 or GM4 for installation up to 100 bar and where the metering pressure is greater than 21 mbar. |

### 4.4.2 Multi-stream systems

Multi-stream systems may be installed for a variety of commercial and technical reasons, including:

**(a)** continuity of supply;

**(b)** partial continuity of supply for essential loads;

**(c)** improved pressure control over the required flow range;

**(d)** achieving the required capacity;

**(e)** provision of standby metering; or

**(f)** provision of reference metering.

### 4.4.3 Pressure control

If pressure control is required the following design elements need to be taken into account:

**(a)** size of the PRI/regulator;

**(b)** regulator lock-up;

**(c)** response time;

**(d)** pressure set points for twin stream installations;

**(e)** monitor regulators;

**(f)** auxiliary systems;

**(g)** slam-shut valve (SSV) systems; and

**(h)** pressure (creep) relief.

## 4.5 Commissioning

Prior to commissioning, the installation must be tested in accordance with the relevant version of IGEM/UP/1, which covers strength testing, tightness testing and purging of gas installations. This is to ensure that the installation does not leak and that there is no extraneous air in the pipework. Once the installation is tested and purged the meters can then be commissioned in line with the manufacturer's instructions and the appropriate standard as per Table 4.2.

## 4.6 Hazardous area classification

To limit the risk of explosion, care must be taken when planning to make electrical connections to gas metering equipment. The Dangerous Substances and Explosive Atmospheres Regulations (DSEAR) came into force in December 2002. Details of how to comply with regulations are covered in IGEM standard IGEM/UP/16 *Hazardous Area Classification for Natural Gas Installations Downstream of Primary Meter Installations*.

© The Institution of Engineering and Technology

# Water metering

## 5.1 Introduction

This section provides a summary of the steps for selecting a meter to monitor the supply of water to a site, part of a site or individual facility (often referred to as sub-metering). It describes the selection process and the main types of meter available.

Metering and sub-metering of water may be carried out for one or more of the following purposes:

**(a)** providing a cross-check on the water provider's meter;

**(b)** as part of a water awareness or reduction programme;

**(c)** cost/resource allocation between different buildings, processes or users;

**(d)** billing tenants;

**(e)** identification of leakage and water losses across a site;

**(f)** checking for unauthorised uses (for example, from hydrants); and

**(g)** enhancing a company's green credentials and/or compliance with ISO 14001 environmental practices to demonstrate good management of water resources.

Water and energy metering can be viewed as complementary activities to identifying inefficiencies and improving resource management given that, for many sites, a significant amount of energy is used to heat water.

## 5.2 Legal metrology with respect to water meters

The only formal legal metrology with respect to water metering is that meters used by water companies supplying water to household premises must meet the requirements of the MID. No such requirements exist for non-household customers. However, the codes for the opening of the water retail market to non-household customers in 2017 are still being developed. This means that, from 2017, non-household customers will not be tied to their local water company for water and wastewater services but may procure those services from other companies. Within the Market Architecture Plan (MAP), which is currently on version 4 (MAP4), there is reference to BS EN ISO 4064:2014, which is the standard used for demonstrating compliance to the MID. This means that, from 2017, meters used for billing water to non-household customers will also be required to demonstrate compliance with the MID.

Other regulated instances include:

**(a)** meters used by water companies supplying water to non-household premises are currently covered only by general weights and measures legislation. There is no explicit statutory requirement to use MID-approved meters. However, the codes for the open retail market for water supplied to non-household customers, which will be operational from April 2017, make reference to BS EN ISO 4064:2014. This

© The Institution of Engineering and Technology

standard encompasses all the requirements of the MID and compliance with the standard can be used to demonstrate compliance with the MID.

**(b)** some water abstraction permits may include specific requirements for meters or metering practices.

**(c)** meters used for measuring discharges to the environment in England under EPR permits must meet the requirements contained in the Environment Agency standard for the self-monitoring of flow. This includes a requirement that all new or replacement meters used for this purpose shall have been certified as compliant with the MCERTS *Standard for Continuous Water Monitors Part 3 Water Flowmeters*.

## 5.2.1 Summary checklist for meter selection:

The following list summarises some of the key considerations when selecting a meter.

### Application

- What is the purpose of measurement/monitoring?
- What is needed – volume totals/sub-totals or rate?
- Temporary or permanent installation?

### Performance

- What performance aspects are important (for example, accuracy or repeatability for trends)?
- What level of performance is required?
- How will data be collected (for example, visual reads, pulse connection to logger, direct to Scada/BMS)?
- How often will data be required?
- What range of flow rates are to be measured?
- Are there any limitations on head loss?

### Fluid

- What is being measured, for example, clean water, wastewater, recycled water, raw water, chemically dosed water?
- Has compatibility with the meter been confirmed?
- What is the water temperature?

### Installation

- Is enough space available?
- Uni- or bidirectional flows?
- What is the pressure?
- What will be the means of connection into the pipework?
- Hygienic application?
- Is there adequate availability of power?

### Environmental

- Is the meter likely to be submerged or buried?
- Will installation be undertaken in contaminated land?
- Are there any sources of electrical interference close by?

© The Institution of Engineering and Technology

**Economic factors**

- What is the expected life of the meter?
- What are the likely maintenance requirements?
- What is the available budget?

# 5.3  Meter selection

There are two important aspects that must be considered in order to obtain data about reliable water consumption: the most appropriate type of meter to use and the size of the meter. To choose the most appropriate type, the following factors should be considered:

**(a)** performance and facilities required;
**(b)** properties of the fluid;
**(c)** installation;
**(d)** environmental considerations; and
**(e)** economic constraints.

## 5.3.1 Performance and facilities

### 5.3.1.1 Accuracy and repeatability

For general water monitoring, meters produced to the current international standards for water meters (BS EN ISO 4064:2014, OIML R-49 or BS EN 14154:2005) should be used. These standards require a maximum ± 5 % error at lower flow ranges and ± 2% at medium to high flows, though actual measurement error is usually much less than this. The first two standards listed also have a higher accuracy specification (± 3 % at low flows and ± 1 % at medium to higher flows) for where higher accuracy metering is required.

Identification of leakage is frequently carried out by the analysis of trends over time, particularly the examination of flows at periods of minimum demand. Repeatability at low flows is therefore an important consideration when selecting a meter for this purpose.

### 5.3.1.2 Flow range

As meters of different types have different operating ranges and accuracy is a function of flow rate, selecting the right sized meter is crucial to the flows it can record within an acceptable error. Meters should be sized on the rates of flow that the meter will be required to measure, not simply on the bore of the adjoining pipework. Often the meter required will have a smaller nominal bore than the pipe into which it is fitted as pipes are frequently over-sized to reduce head loss or allow for future demand growth. An over-sized meter may be working at the low end of its range where performance will be generally poorer than in the middle of its range.

However, compromise may be necessary to balance other factors, for example, reducing the pipe so that a meter with a smaller bore than the pipe can be installed will add cost to the installation and introduce an additional head loss. Also, if the range of flows is very wide and a small mechanical meter is fitted to capture low flows, the meter may become damaged if it then runs for long periods at very high flows.

Meters produced to one of the standards listed above will have their operating range defined by a nominal flow rate (Q3) and a turndown value (R).

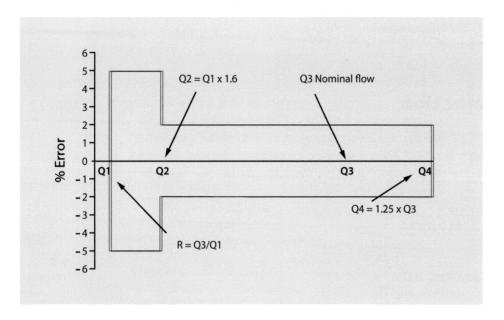

### 5.3.1.3 Parameter required

Decide whether total consumption or a flow rate is required. Most electronic meters will give both a flow rate (the amount of water flowing through the meter at any point in time) and a total consumption value. Mechanical meters will give a total consumption value and, where flow rate is required, this must be deduced by further processing of, for example, a pulse output signal.

### 5.3.1.4 Outputs

In addition to a visual register, most meters are available with different output options for connection to a data collection system.

▼ **Table 5.1**    Water meter output options for connection to a data collection system

| Output type | Description |
|---|---|
| Pulse output | A pulse is generated every time a given volume passes through the meter (e.g. 1 litre, 1 m³ etc.). For mechanical meters, the pulse value is usually fixed from a limited number of options available when the meter is purchased. Electronic meters can usually be programmed by the user for whatever pulse value is most appropriate. The total number of pulses gives the overall consumption. |
| Analogue output | An electrical signal, e.g. 4-20 mA current or 0-5 V, is generated in proportion to the flow rate through the meter. Electrical analogue outputs give flow rates that must then be integrated to give total consumption. Analogue outputs are not generally available from wholly mechanical meters as electrical power is required to generate them. |
| Digital outputs | Most electronic flow meters can be supplied with digital outputs, e.g. Hart, Profibus, MBus. |

© The Institution of Engineering and Technology

| Output type | Description |
|---|---|
| Remote communications | Some meters incorporate radio, GPRS or GSM modules to communicate. This can save on cabling, but additional hardware and software to capture and interpret the signals are required. Some suppliers will offer a service where the data is sent to a server and can be accessed through the web. |
| Alarms | Most electronic meters incorporate user configurable alarms, e.g. alarms that may indicate a burst pipe on systems with high flows. The number and type of alarms required should be considered and specified at the time of order. |

## 5.3.2 Fluid properties

All the meter types described in Section 3 are suitable for cold potable water or clean water from storage tanks. Where the water has been chemically dosed then the materials in contact with the water should be checked for compatibility. Smaller mechanical meters are available in different temperature ranges, with specific versions for hot water. Meters produced to one of the water meter standards (see above) will have a temperature rating indicated by a 'T' followed by a number. For example, a meter designated as T30 will be rated for water up to 30 °C.

If water has been taken directly from a surface or groundwater source, or is recycled greywater, then consideration should be given to any solids that the water might contain. Mechanical meters are generally less suited for water containing particulate or other debris. Fine particulate may pass through a mechanical meter but is likely to accelerate wear of the meter and lead to degradation of accuracy and early replacement. There are some mechanical meters designed specifically for irrigation systems but these will generally have a poorer specified accuracy than a potable water meter. Similar considerations arise with meters for wastewater.

## 5.3.3 Installation considerations

The quality of installation and installation conditions can significantly affect the in-service performance of any type of meter. The meters described in Section 3 are designed for pressurised pipes running full.

▼ **Table 5.2**   Considerations for selecting a water meter

| Consideration | Description |
|---|---|
| Flow direction | Consider whether flow will always be in the same direction through the meter or whether backflow will also occur. Most mechanical meters will pass reverse flow but, for many, performance will be poorer and prolonged reverse flows can lead to damage. The total registered volume will also decrease. Electronic meters can usually be programmed to provide separate totalizers for forward, reverse or net flows. |
| Meter orientation | Some types of meter, notably jet types and combination meters, perform better in horizontal pipes. Other types of meter are suitable for horizontal, vertical or sloping pipelines. For vertical installation, flow should be upwards through the meter. |
| Adjacent pipework | Manufacturers will provide installation guidance for individual meters. Such guidance will specify the required straight lengths of pipework upstream and downstream. These need to be free of disturbances (such as bends and partially shut valves). The space available to install the meter should be checked. |

© The Institution of Engineering and Technology

| Consideration | Description |
|---|---|
| Method of installation | The majority of the meters described in Section 3 attach to the pipework by flanged connections or threaded connections (the latter typically for smaller sizes up to 40 mm nominal bore). Clamp-on ultrasonic meters can be installed directly on the outside of a pipe, which may be of use for large pipes or lines that are difficult to shut down. This can be for permanent or temporary installations. Accurate knowledge of parameters, such as wall thickness, tuberculation levels and material type, is critical to the installed accuracy of the meter. The performance of clamp-on meters is heavily dependent on the quality of the installation and will not generally be as good as for a spool piece meter. |
| Availability of power | The availability, or lack, of a local power supply may limit meter choice. Mechanical meters do not require any electrical power supply, drawing the energy needed to drive the meter from the flow itself. Other technology types can be specified with long-life batteries or with the facility to be powered by renewable resources (e.g. solar). |

## 5.3.4 Environmental considerations

### 5.3.4.1 Water or soil ingress protection

Many meters will need to be installed in pits or chambers liable to soil ingress, insect occupation or flooding. Any electrical equipment should be protected from the ingress of dust or water that may impair its performance. The degree of protection for electrical equipment is specified by its IP code. Full explanations of the IP code can be found in IEC 60529 *Degrees of protection provided by enclosures (IP Code)*.

Any electrical enclosure (such as a meter sensor or datalogger) that is likely to be in a flooded chamber should be protected in accordance with IP68, i.e. is submersible, to an appropriate depth for extended periods, possibly indefinitely in certain cases. Electronics units, including dataloggers or telemetry equipment installed in cabinets above the ground, should also be protected, preferably to IP65. Even though they may not be subject to the direct impact of water, condensation can be a problem.

### 5.3.4.2 Installation in contaminated land

There are two areas of concern if meters need to be installed in contaminated land:

**(a)** diffusion of pollutants through the meter body leading to contamination of the water supply; and

**(b)** corrosion or weakening of the meter body or other components (such as seals) leading to failure or contamination.

There is a Water Industry Specification (WIS 4-32-19: *Water industry specification, Polyethylene pressure pipe systems with an aluminium barrier layer for potable water supply in contaminated land – Size 25 mm to 630 mm*) for pipe systems in contaminated land and, although meters are not specifically mentioned, they could be considered. Most manufacturers of polymer bodied meters have had their meters tested in accordance with this specification.

© The Institution of Engineering and Technology

### 5.3.4.3 Electrical interference

If there are major sources of electrical interference, for example, pumps or switchgear, close to the proposed meter installation point this needs to be recognised when specifying the meter so that any necessary special screening arrangements can be included.

### 5.3.4.4 Economic factors

As with any equipment, economic considerations are important so that the whole life costs can be taken into account. Such considerations include purchase price, installation costs, maintenance costs and meter life.

Mechanical meters will be prone to wear and will require periodic replacement, the appropriate interval for which will vary depending on the duty of the meter.

For electromagnetic meters, again, there is little routine maintenance that can be done. All meters should be inspected regularly to ensure that they are still working, have not been damaged and there is no evident corrosion or other significant deterioration to key parts. In addition the integrity of the electronic systems can be checked using software supplied by the meter manufacturer. This is known as electronic verification.

Clamp-on ultrasonic meters may require the coupling compound to be renewed periodically. The frequency depends on the environment – hot, dry environments will lead to the compound drying out and giving poor coupling between the transducer and the pipe more quickly than damp, cool environments, such as meter chambers.

Mechanical meters will tend to wear over time leading to poor flow accuracy. Research by Water Research Centre (WRc) for UKWIR suggests a service life of 12 years for Woltmann turbine meters. For rotary piston meters, service life is typically 15 years.

# 5.4 Meter types

## 5.4.1 General

The principal types of meter for water metering are summarised in this section.

The mechanical meters require no external power supply and typically show the total amount of water consumed on an integral register. They are better suited to clean water although some particulate will pass. They will tend to jam or under-read if any fibrous or filamentous material is present.

Electronic meters are more versatile, although require some power, whether from battery or mains. They can provide a direct measurement of flow rate and are suitable for supplies containing solids, such as particulate etc.

The electronic meters tend to be more expensive than mechanical meters.

## 5.4.2 Mechanical meters

### (a) Rotary piston meters
These are the most commonly used type of meter for household or small commercial revenue metering. These meters can measure very low flows and are suitable for metering supplies where there are storage tanks with ballcock-type float valves that gradually close as the float rises. They can be installed in any orientation and are one of the few types of meter that require no straight upstream or downstream pipe lengths. This makes them

suitable where space is limited. They record the total volume passed and can be fitted with pulse outputs. Some modern versions have electronic, rather than the traditional mechanical, registers. Particulate can accelerate wear or in severe cases cause jamming.

### (b) Woltmann turbine meters

These are widely used for revenue metering on industrial premises, typically for supplies from 50 mm to 150 mm nominal bore. They usually require 5 to 10 diameters of straight pipework upstream. They record the total volume passed and can be fitted with pulse outputs. Some modern versions have electronic, rather than the traditional mechanical, registers. They will pass some particulate, although this may accelerate wear of the meter. They are not suited to water that contains weed or other fibrous material.

### (c) Combination meters

Such meters combine a turbine meter with a small piston or jet meter in parallel with a valve that diverts high flows through the turbine and low flows through the small bypass meter. They can be used where there is a very high flow range, for example, a building that is fully occupied and operational in the day but is not used much at night. They require at least 10 diameter straight pipework upstream and are only suited to clean water.

### (d) Jet meters

Single and multi-jet meters are available from 15 mm to 100 mm nominal bore and are widely used for sub-metering. Generally they do not have the low flow sensitivity of a rotary piston meter but are suited to monitoring consumption on supplies that are directly fed (i.e. straight from mains or pumped). Many models can be fitted with a pulse output.

## 5.4.3 Electronic meters

### (a) Electromagnetic meters

Electromagnetic meters are available in battery or mains powered versions across a wide range of sizes, from a few millimetres to 3 meters nominal bore. They have long been the mainstay for many process monitoring applications and they are becoming widely used in water network and revenue metering. They have a clear bore through the meter so can pass solids etc. On some models the bore is profiled to increase sensitivity at low flow rates.

### (b) Ultrasonic meters

Ultrasonic meters share similar advantages to electromagnetic meters. However, they are available both as spool-piece-type meters and also clamp-on versions. The latter versions can be attached to the outside of the pipe. This makes them useful for temporary or survey-type applications as there is no need to cut into the pipe. They can also be fitted to pipes that cannot be shut off or where high integrity is required. Clamp-on meters require good installation conditions, typically 10 diameter straight upstream pipe or more if there is a pump or other severe source of flow disturbance upstream. Accuracy is highly dependent on reliable information on the pipe material and dimensions.

## 5.4.4 Other types

There are many other types of meter that can be used for measuring water flows although these would not usually be used for general consumption monitoring or sub-metering.

© The Institution of Engineering and Technology

▼ **Table 5.3** Additional meter types for measuring water flows

| Meter type | Description |
|---|---|
| Differential pressure (orifice plate, Venturi etc.) | These are rarely installed for water applications now. The need for a primary device (e.g. orifice or Dall tube) plus a differential pressure cell makes installation relatively more complicated. |
| Coriolis meters | These are intended for mass flow measurement where a high degree of precision is required, such as process control and fiscal measurements of high value commodities. Their relatively high cost and limited size range are unlikely to make them suitable for general consumption monitoring or sub-metering. |
| Variable area/float meters | These provide a visual indication of flow rate. Not suitable where total consumption over time is required. |
| Vortex meters | These are mainly used for gas and steam. Limitations on minimum flows and turndown are likely to make them unsuitable for water consumption. |
| Thermal meters | These tend to be used for gas flows. |
| Positive displacement meters | Rotary piston meters, a sub-group of positive displacement devices, have been included but other meters in the group have not as they are more suitable for fuel or process flows. |

In addition, other specialised meter types are available for flows through open channels or partially filled pipes and sewers.

## 5.5 Regulated applications

Where sites are subject to EPR permits or where raw water is being abstracted from a ground or surface water source, the following Environment Agency documents provide further guidance on what is acceptable:

**(a)** Abstraction: *Abstraction Metering Good Practice Manual Environment* Agency Report W84 (see https://www.gov.uk/guidance/water-abstraction-how-to-make-sure-your-meter-is-accurate)

**(b)** Discharge: *Minimum Requirements for the Self Monitoring of Flow* (see www.mcerts.net for the current version)

© The Institution of Engineering and Technology

# Heat metering

## 6.1  Introduction

Heat metering has been used for many years to measure and charge for the consumption of heating and cooling. Heat metering requirements were mostly unregulated for many years up until the inclusion of the subject in the Building Regulations. With the introduction of the Heat Network (Metering and Billing) Regulations 2014, heat metering has become a legal requirement for both end-point and bulk metering on a heat network or communal networks.

A heat meter is a piece of equipment that is fitted into a closed loop system (see Figure 6.1) and measures the amount of energy that is consumed by a load, for example, the heating of a swimming pool or residential properties. The type of meter and required outputs should be considered carefully to make sure that the meter meets the specification for both media type and end user requirements.

▼ **Figure 6.1**    Typical heat metering layout

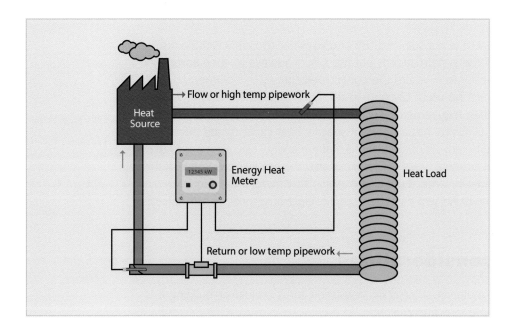

## 6.2  Basic heat meter

All heat meters, irrespective of manufacture and flow measurement principles, will have the same basic components:

**(a)** a flow/volume measuring device;
**(b)** a pair of matched temperature sensors; and
**(c)** an energy calculator.

© The Institution of Engineering and Technology

### 6.2.1 Flow/volume device

There are three main types of flow measurement:

**(a)** mechanical turbine;
**(b)** ultrasonic; and
**(c)** electromagnetic.

The flow/volume device measures any water flow rate passing through the system load. At the same time, the meter also measures the temperature difference (known as Delta T) between the flow (hotter pipe going to the system load) and the return (colder pipe going away from the system load) by using the temperature sensors (see Figure 6.1).

### 6.2.2 Energy calculator

Using this information, the energy calculator records the volume of water and the temperature difference over a period. The energy calculator then adds in a correction factor. This factor is generated by using the temperature of the flow and return water and then setting or correcting the density and specific heat of the liquid in the energy calculation. The energy calculator has to be programmed correctly for the line that the flow meter is measuring the liquid in i.e. high or low temperature line. Once the corrector has applied this correction, the calculator then displays the energy consumption.

If there is no heat being consumed from the system (flow and return temperatures are the same so there is no temperature difference) and there is still flow, the meter will record no energy use. Likewise, if there is no flow but there is a temperature difference, again, the meter will record no energy use.

If the medium in the system is not standard hot or cold water (possibly with corrosion inhibitor for protection of the system added) but has Glycol additive in, then an energy calculator that has the ability to measure the correct percentage, by volume and type of Glycol in the water, must be used. Glycol is generally an anti-freeze used in various systems dependent on their requirements, such as a solar heat system or certain Heat and Power (CHP) systems. Glycol has a different specific heat capacity and density to standard water so has a different set of calculations that the energy calculator has to use. It is also worth noting that different types of Glycol can alter the acoustic properties of the liquid and thus cause issues in some, if not all, ultrasonic flow meters. Consequently, an alternative type of flow measurement must be used, i.e. electromagnetic or mechanical.

## 6.3 Design considerations

When designing a heat metering strategy or installation for a project or building it is important to consider some selection criteria:

**(a) Specification**
What does the project specification say, i.e. what is the flow measurement principle? What certification level is required? What is the life expectancy of the equipment?

**(b) Performance rating of the meter**
Does the meter work at the temperature and pressure of the system media? Can the meter work at the design flow rates?

© The Institution of Engineering and Technology

### (c) Meter accuracy

Does the meter meet with the required accuracy class either for the project specification/criteria, EN1434 or MID class 2/3? For example, RHI requires a MID Class 2 meter to obtain RHI eligibility. If fiscal billing is being used for selling heat to a resident or commercial tenant, then has the accuracy been meet for this function?

### (d) Legal requirements

Is there a legislative requirement for any taxation redemption or planning requirements, i.e. RHI, CHPQA and Building Regulations for communal buildings?

### (e) Data collection methods

Does the equipment give the required outputs to a building management system (BMS) or to a currently installed automatic meter reading (AMR) system or any other specified system? Is it open protocol (SMART systems)?

### (f) Location of meter installation

Can the device be installed in the required location and pipe work plane? Can the device be programmed to work in either flow or return?

### (g) Service and maintenance requirements

What are the service and maintenance requirements? If the meter requires yearly verification then what access to the equipment is required? Who has to carry out the inspection?

## 6.4 Meter point installation

### 6.4.1 General

The location of the installation is very important. If the meter is installed in the wrong place you may not be able to measure the desired heat load consumption. The flow meter and the temperature sensors have to be placed in locations that ensure all of the flow and temperature differences or Delta T are measured in accordance with the specification of the project. For example, if you are required to measure the heat energy consumption of a block of apartments, then the flow meter and temperature sensors must be installed in the main riser at the first point of entry before any branches off to the lateral corridors.

Access will be required to manually read the heat meter or to carry out any service and maintenance. For example, if a meter is to be positioned in a ceiling void then an access hatch must be positioned by the meter to allow access.

The physical installation position of the flow measuring device is also very important as each measuring principle and manufacturer has different upstream and downstream unrestricted straight pipework runs, normally designated in a number of diameters of the pipe line i.e. 5 × diameter. If the pipe is 100 mm in diameter, then 500 mm of unrestricted clear straight pipework is required before the flow reaches the measuring device. This would have to be checked for each meter type. The mounting or installation of the flow meter, as to which pipe line it is installed in, be it the flow (high temperature) or return (low temperature), must be the same as the energy calculator set up and is crucial to the accuracy of the heat meter.

### 6.4.2 Location, positioning and meter mounting

Where is the flow meter mounted in the pipework and have the correct number of straight lengths been allocated? To avoid air being trapped in the meter, do not install at the highest points. If mounted in a vertical pipe consider whether there is always water in that section of the pipe, as this is vital for correct measurement and accuracy. Is the meter set up correctly for the flow/return and can this be changed on the meter? Is there access for routine maintenance? Can the meter be seen for manual reading purposes?

### 6.4.3 Flow rates and temperature

Does the meter (flow-measuring unit) proposed for the project work within the desired flow range of the system criteria? Each and every flow meter will have a minimum and maximum flow rate, which will be unique to each manufacturer and measuring principle. In addition, the temperature of the medium being measured should be considered, as the display may have to be remotely mounted to avoid excessive heat damage to the electronic energy calculator, or the flow measuring chamber may not be capable of working with the medium temperature.

### 6.4.4 Sensors/ancillary units

The temperature sensors of a heat meter are produced in matched pairs and they are designed to read the temperature difference. They must only be installed or changed in matched pairs to maintain the accuracy and integrity of the overall heat meter. Sensors can be inserted either directly into the water or into dry pockets or 'wells'. The tip of the sensor should be positioned at 50 % of the pipe diameter (see Figure 6.2). If the cables on the sensors are to be extended then the cable type and extended length must be of the same to avoid adding any additional resistance. There are many types of ancillary equipment/option cards that vary from manufacturer to manufacturer and these are bespoke to each type of meter. You must make sure that the meter purchased/installed has the correct option for the desired output/function and meets both the specification and client needs.

▼ **Figure 6.2**   Ideal position of sensor pocket

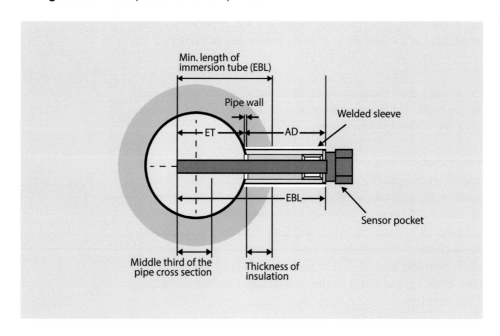

© The Institution of Engineering and Technology

### 6.4.5 Power supply

Power supply requirements depend on the type of meter. Electromagnetic-type meters will require a supplied voltage (generally 24 V AC/230 V DC). Ultrasonic meters can come in both mains 230 V AC and battery versions, with varying battery lives. Please note that a certain supply may be required, depending on the manufacturer and the option cards fitted/required, i.e. an analogue output card may require an external 24 V DC supply but the meter itself only requires a 16-year battery – check with the supplier/manufacturer.

**Note:** With most modern day electronic heat meters the set-up data and totalised readings will be held in a non-volatile memory in the event of a power loss. No readings will take place at the time of power loss but will start reading when power resumes.

### 6.4.6 Labelling/identification for use and information

It is best practice to have a unique reference to each and every meter installed in a project and to keep to the same labelling structure for the entire project. This is called a meter point reference (MPR). For example, a meter point label may say – MPR 0001 HM 1. This is the meter location reference, not the meter, as in time the meter itself may have to be replaced but the required location metering will still be relevant.

The list of MPRs should be handed over to the end client and maintained with the type of information outlined in Table 6.1.

▼ **Table 6.1**    Typical heat meter handover information

| MPR Number | Service | Location | Meter type and size | Serial number |
|---|---|---|---|---|
| 0001 HM1 | Heating | X ray block C | UH50 65 mm | 1234567 |

### 6.4.7 Inspection and testing

As the meters installed may be used either for monitoring only or for fiscal purposes, it is very important to organise and implement a structured verification/calibration regime that fits with the client's needs, manufacturer's guidelines and current legislation. For monitoring, a very basic regime may be required every 2-4 years. For fiscal metering a more structured and rigorous regime will be required, especially if taxation rebates are being claimed. This could be on-site inspection/verification every year with a meter removal and off-site calibration check every 5 years. Each manufacturer may also offer extended warranty if the meter is verified every year!

### 6.4.8 Commissioning

The installation of any heat meter should be carried out by a competent person who has the correct tools and software to set the meter correctly (usually using either a metering engineer (from the manufacturers) or a controls engineer).

By commissioning the installation of the meter, it is ensured to be fitted correctly (straight lengths, mounting position etc.). The start readings would be logged along with all the meter data, any set-up information and any bespoke output settings required by the client.

© The Institution of Engineering and Technology

# SECTION 7

## Meter data and communications

### 7.1 General

It has been possible to collect data from utility meters for many years. Traditionally, this was done through the use of a 'read' switch that was fitted in the meter and would pulse in line with a cycle of the meter index. This pulse would represent a known value, for example, 1 unit of electricity or $^1/_{10}$ m$^3$ of gas. These pulses were then counted and stored in a datalogger. The data could then be periodically downloaded and used as required.

This method is still widely used in the non-domestic gas and water market. These days, however, the use of solid state metering has allowed the development of much more complex meter functionality and, in particular, two-way remote communications for readings, as well as control functions. There is no overall standard for the communication of data with these meters, but the following is a guide to making the best use of the data available.

### 7.2 Communications interfaces

Figure 7.1, as adapted from CEN/CENELEC/ETSI TR 50572, is an example of an advanced metering system and shows the interfaces for a smart metering system. The blue interface symbols are all supported by standards and the same diagram can be used to describe a sub-metering system that is intended to supply data to a BMS system, which, in the diagram, is shown as the Automatic meter reading (AMR) head end.

The meter may support a port dedicated to a remote display, shown as H1.

It is likely that the meter will communicate within the building through a local area network and there will be an interface ('M') with the local network access point (LNAP). Other devices for automation purposes may also be connected to the LNAP through the H2 interface. (This may be KNX, Modbus, BACnet etc.)

In the case of the Great Britain smart meter system, the LNAP is the communications hub and could also be the consumer access device (CAD). The CAD can connect to the meter via the M interface, which is security protected, but the consumer can access the data via the user-defined H2 interface.

Figure 7.1 illustrates an example of meters communicating over a neighbourhood network (such as a powerline) to an access point in the substation (NNAP) over the 'C' interface.

The 'G1' interface reflects a meter that communicates over a wide area network (such as GPRS) directly with the head end system.

© The Institution of Engineering and Technology

The designer of the sub-metering system should take account of the interfaces in the system to ensure that the metering interfaces are compatible. TR50572 provides guidance as to the standards available for fiscal smart metering interfaces.

More information on smart metering is available in Annex D.

▼ **Figure 7.1**    Communications interfaces in a metering system

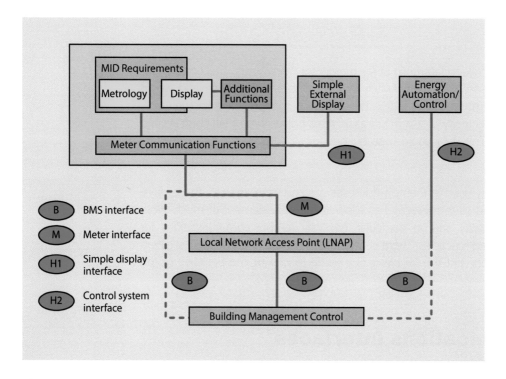

The system shown in Figure 7.1 represents a metering and control system. The interfaces between each of the component parts are represented by the blue circles. These interfaces must be considered when designing a BMS.

Meters are shown in the top left; the same basic model applies to any meter for any fuel type. The meter will have some type of communication port represented by the M interface. In some cases the meter will have the functionality to interface with a remote display device and if so will be fitted with an H1 interface. (This may be the same as the M interface.)

Communications between the building control system will normally be via an LNAP, which is fitted with an M interface on the metering side and a B interface on the building control system side.

Building controllers, such as temperature sensors, valves, etc., may also be connected via the LNAP through the H2 interface.

An alternative architecture may be built where the building control system interfaces directly with the metering and/or the controllers. In this case, the metering and controllers must have a B interface or the building control system must support the M and/or H2 interfaces.

© The Institution of Engineering and Technology

| Interface | Electricity | Gas | Heat | Control |
|---|---|---|---|---|
| M | IEC 62056 Series (DLMS) via:<br><br>• Optical port<br>• RS 485<br>• S0 open collector pulse output<br><br>M-Bus (EN 13757)<br><br>Modbus<br><br>ZigBee SE<br><br>TCP/IP via RJ45<br><br>20 mA loop | R5 Pulse output<br><br>IEC 62056-21 Optical port<br><br>M-Bus (EN 13757):<br><br>• Wired<br>• Wireless<br><br>ZigBee SE | IEC 62056-21 Optical port<br><br>M-Bus (EN 13757):<br><br>• Wired<br>• Wireless<br><br>ZigBee SE | |
| H1 | IEC 62056-7-5<br><br>S0 serial output | | | |
| H2 | | | | Modbus<br><br>EN 50090 (KNX)<br><br>ZigBee SE<br><br>TCP/IP<br><br>WiFi<br><br>Powerline |
| B | | | | Modbus<br><br>EN 50090 (KNX)<br><br>ZigBee SE<br><br>TCP/IP<br><br>WiFi<br><br>Powerline |

Table 7.1 provides a list of some of the standards that may be used for the system interfaces. The list is not exhaustive but shows those most commonly used in the industry.

Further information on this architecture can be found in CEN/CENELEC/ETSI TR 50572 (ftp://ftp.cen.eu/cen/Sectors/List/Measurement/Smartmeters/CENCLCETSI_TR50572.pdf).

# 7.3  Use of pulse outputs and dataloggers

Consumption is monitored using pulse technology. A simple two-wire low voltage cable transmits the position of a switch, which generates a pulse that corresponds with consumption. Sometimes pulses are transmitted using a proprietary low-power radio protocol.

Pulses from one or more meters are collected in a datalogger-modem or communications hub and processed according to unit and value, stored usually in time intervals of 30 minutes (electricity) or one hour (gas and water). The communications hub is then accessed through a point-to-point network (GSM, GPRS, SMS or IP), usually using a protocol that is proprietary to the vendor. Most mechanical meters can be provided with a pulse output, and there are many pulse counting devices available in the industry.

Care must be taken to ensure that the register in the pulse counter matches that of the mechanical register. Missing pulses and incorrect multiplication factors often cause errors in reading. Systems are configured to either 'dial-in' or 'dial-out'. Most meters are dial-in, in other words, the interrogating station initiates the call to retrieve data. However, some meters dial-out at pre-set times during the day or month. This is to conserve power, or to provide better compatibility with business network security. It is much better to install meter 'clients' on a corporate network than meter 'servers'.

The disadvantage of dial-out is that the in-station is more complicated, since it needs to listen for calls. Often, the in-station is provided as a service by the meter vendor, who saves the data in flat data files using the internet.

It is important when choosing a metering system to know what the license conditions are for the in-station software, or 'head end'. Most vendors include the software with the order for meters, but the license may be restricted to the number of sites and may only include support for a limited period. If the reading service is provided by the meter vendor or agent, then the terms of engagement must be established for the expected life of the meter. For larger systems, and if you plan to build your own data retrieval system, it is advisable to ask the vendor for access to documentation and support for the protocol at the same time that the meters are purchased.

# 7.4  Common meter protocols

## 7.4.1 Categories of meter protocol

Common meter protocols fall into two categories: electricity metering and battery powered meters.

### (a) Electricity meter protocols
Electricity meter protocols follow either IEC or ANSI standards. IEC covers most of Europe and the rest of the world, while ANSI is concentrated in the USA and some Scandinavian countries.

The IEC 62056 series of protocol standards cover DLMS/COSEM and defines the data objects and methods for reading and writing data to meters. It provides the details for how to construct messages to a common format but does not define the way it is implemented. Although the messages are constructed to a common standard, their format may differ between manufacturers.

### (b) Battery powered meter protocols
BS EN 13757 covers the M-Bus protocol used for communications to battery powered meters. This is much simpler than DLMS/COSEM, with shorter messages requiring less processing overhead and hence less battery power. The protocol defines the data items and how to read/write them.

© The Institution of Engineering and Technology

## 7.4.2 Communications protocols for sub-metering

For sub-metering, a range of communications protocols are available, including:

### (a) BACnet

BACnet is a communications protocol for building automation and control networks. It is covered by ISO 16484-5.

BACnet was designed to allow communication of building automation control systems for applications such as heating, ventilation and air-conditioning (HVAC) control, lighting control, access control, and fire detection systems and their associated equipment. The BACnet protocol provides mechanisms for computerised building automation devices to exchange information, regardless of the particular building service they perform.

The protocol specifies 54 standard object types. Meter manufacturers provide object mapping of custom Modbus addresses so that an extended range of metering specific objects can be accessed.

### (b) KNX

KNX is a European protocol standard (EN 50090) used for building management systems. The protocol is widely used throughout Europe.

There are a number of physical layers supported:

**(i)** twisted pair;
**(ii)** powerline;
**(iii)** radio;
**(iv)** infrared; and
**(v)** Ethernet.

Meters are available that support a KNX interface for use within KNX-based BMS.

### (c) M-Bus

M-Bus (meter bus) is probably the most widely used metering communications protocol with regard to data exchange and interfaces and is based upon the European standard for heat meters, namely EN 13757-2/EN 13757-23. It was originally developed by a German consortium in the 1980s to enable reliable automatic meter reading primarily for heat but as M-Bus is an open protocol, information to read and write to M-Bus slaves is available to anyone. This in turn means that M-Bus has been adopted by manufacturers of heat and cooling devices, electricity, water and gas meters.

M-Bus is available in meter types as follows:

**(i)** water – hot and cold/potable.
**(ii)** gas – various.
**(iii)** electric – single and 3-phase.
**(iv)** heat – various.
**(v)** pulse to M-Bus convertors.

M-Bus is an inexpensive solution for widespread networking and remote meter reading for large numbers of heat, gas, water, and energy meters from different manufacturers. The task of reading small amounts of data per energy consumption places only minimal demands on transmission speed (2,400 baud) and allows for very long transmission distances (+3 km) using a simple twin-core non-shielded standard cable. M-Bus allows 250 meters or 'nodes' on one 'loop'.

### (d) Modbus

Modbus is a serial communications protocol published by Modicon for use with programmable logic controllers.

It was developed for industrial applications, connecting devices such as sensors to a remote terminal unit in SCADA systems.

The protocol can be used over numerous physical layers including a version for TCP/IP with a dedicated port (502).

Meters that support Modbus to allow consumption and other data to be read over the network are available.

### (e) Z-Wave

Z-Wave is a radio-based system designed for building control systems using the 868 MHz ISM band. It supports a wide range of control devices and can support the transportation of metering data.

Meters are available that support Z-Wave.

### (f) ZigBee

ZigBee Smart Energy has been adopted in some countries including the USA and UK for use in metering applications. This defines data commands and attributes associated with metering applications and how to access them. The protocol defines a mandatory set of commands and attributes that must be implemented to gain ZigBee certification. This mandatory set allows a very basic smart metering application to be built. There are many optional commands and attributes that can be specified for more advanced applications and care must be taken to ensure that these are included in any system specification prior to procurement.

There must be an associated protocol interface conformance specification (PICS) detailing all the options required for the application that devices must be tested against for the appropriate ZigBee certificate.

## 7.5 Energy dashboards

Metering systems are becoming integrated into energy management systems to provide data for energy dashboards, building control systems and appliance control.

Care must be taken to ensure that the meter data collection system is compatible with other systems on site and that accurate and timely readings can be taken automatically. An energy dashboard is an advanced version of the in-home display (IHD). It is normally a PC monitor, often of large size and displayed prominently on an industrial or commercial site (for example, in the reception area). The dashboard can display usage statistics as well as current demand in units of energy, cost or carbon.

The main purpose of the energy dashboard is to raise awareness about energy consumption, particularly to staff. Much of the reduction in demand can be achieved by staff being more proactive to save energy; an action with a visible response is a good incentive.

Dashboards can take feeders from all utility meters and can also take the temperature inside and outside of the building. It can monitor the number of people in the building

© The Institution of Engineering and Technology

and show that as an 'energy use per person' metric. Often, different areas of the building or departments are grouped and displayed against each other. This provides benchmarking and also catalyses healthy competition for the greenest group.

Dashboards are available as stand-alone devices or, more normally, via an internet provision by a service provider. The latter provides for more updates and messages targeted to the company's usage and circumstances.

▼ **Figure 7.2** Typical Energy Dashboard Display

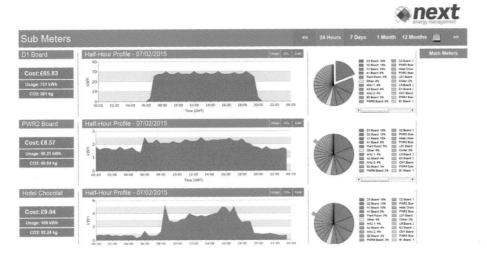

*Courtesy of Next Controls Group*

© The Institution of Engineering and Technology

# APPENDIX A

## Key regulations, standards and guidance

BPF Office ABI Landlord's Energy Statement Guidance and Specification Version 2.3 April 2007

Building Regulations

Building Regulations for England and Wales 2000

Building Regulations (Northern Ireland) 2000

Building Standards (Scotland) Regulations 2000 and Amendments

Carbon Trust *Metering technology overview* (CTV027)

Carbon Trust *Monitoring and targeting* – in depth management guide (CTG008)

CIBSE TM39 *Building Energy Metering*

CIBSE TM46 *Energy Benchmarks*

Confined Spaces Regulations 1997

Construction (Design and Management) Regulations 2015

Construction (Health, Safety and Welfare) Regulations 1996

Construction Products Regulations 2013

Control of Asbestos at Work Regulations 2002

Control of Substances Hazardous to Health Regulations 2002

Dangerous Substances and Explosive Atmospheres Regulations 2002

DECC Enhanced Capital Allowance Energy Technology Criteria Lists (2009/2013)

Electricity Act 1989

Electricity (Approval of Pattern or Construction and Installation and Certification) (Amendment) Regulations (SI 2002/3129)

Electricity at Work Regulations 1989 and Memorandum of Guidance 1989

Gas Act, 1986 as amended

Gas (Calculation of Thermal Energy) Regulations, 1996 as amended

Gas (Meter) Regulations, 1983 as amended

Gas Safety (Installation and Use) Regulations 1998

Gas Safety (Management) Regulations 1996

Management of Health and Safety at Work Regulations 1999

Measuring Instruments (Active Electrical Energy Meters) Regulations (SI 2006/1679)

Measuring Instruments (EC Requirements) (Electrical Energy Meters) Regulations (SI 1995/2607)

Measuring Instruments (EC Requirements) Electrical Energy Meters) (Amendment) Regulations (SI 2002/3082)

Measuring Instruments (Gas Meters) Regulations (SI 2006/2647)

Meters (Approval of Pattern or Construction and Manner of Installation ) Regulations (SI 1998/1565)

Meters (Certification) Regulations (SI 1998/1566)

Pipelines Safety Regulations 1996

Pressure Equipment Regulations 1998

Pressure Systems Safety Regulations 2000

Provision and Use of Work Equipment Regulations 1998

Reporting of Injuries, Diseases and Dangerous Occurrences Regulations 1995

Water Act 2003 (Consequential and Supplementary Provisions) Regulations 2005

Water Supply (Water Fittings) Regulations 1999

   Water Supply (Water Fittings) (Amendment) Regulations 1999

Water Supply (Water Fittings) Regulations (Northern Ireland) 2009

Water Supply (Water Fittings) (Scotland) Byelaws 2014

Workplace (Health, Safety and Welfare) Regulations 1992

© The Institution of Engineering and Technology

# APPENDIX B

# Checklists

## B.1 Metering equipment pre-commissioning checklists

### B.1.1 Electricity

The watch outs for installation/commissioning 'Checklist' of common errors and methods to correct them:

#### (a) Wiring
To measure electric usage you need to measure both voltage and current. It is important to make sure that the wiring is correct when measuring each; even more so when measuring 3-phase electrics.

More often than not, voltage connections and current connections will be bundled together. If not correctly segregated and arranged, a simple phase shift by 120 degrees would cause errors to occur in the metering system. This will then affect billing as well as the understanding of the relationships between waveforms if the metering is being used to measure power quality.

#### (b) Direction
Polarity is important: fitting the CT in the direction of the primary current flow is the first step (P1 & P2) but equally important is the wiring of the secondary connections – see crossed secondary CT wires below.

#### (c) Fusing
Sized, selected and installed by the designer applicable to the metering system being installed.

#### (d) Cable sizing
*see* Burden

#### (e) Balancing
Measure primary and then measure secondary outputs; what goes in must come out and this should be proved and documented in the commissioning process.

#### (f) Wrong phasing
Voltage from one phase L1, is being measured with current from another phase C2. Isolate all the phases and work on one phase at a time to correct the wiring.

#### (g) Crossed secondary CT wires
The problem might be that an incomplete reading or no reading at all is supplied. Check that all CTs have S1 and S2 shown on the output side; if crossed, these will cause current to flow in the wrong direction in the metering and hence a wrong polarity will be provided.

Isolate and work on one phase at a time, correcting any crossed wiring.

### (h) Incorrect ratio setups

Use a hand-held current clamp and compare against the meter reading.

Manufacturers will always declare accuracy and burden levels; always follow manufacturer's instructions when fitting and testing metering systems.

**Note:** Remember that current flowing through any electrical piece of kit will be met with resistance (impedance). Adding up all the devices, including the cabling, will determine the overall burden of the metering system that needs to be overcome to achieve the required accuracy. Manufacturer's charts and guides are available for reference.

## B.1.2 Gas

Detailed guidance on commissioning gas meters can be found in IGEM/GM/8 *Non-domestic meter installations part 3: fabrication, installation, testing and commissioning*.

### Post-commissioning checks

**(a)** after a volume of gas, significant in relation to the capacity of the installation, has passed through the installation, post-commissioning checks should be undertaken. Typically, these should be done between one and three months after the installation has been commissioned but may be undertaken at the first scheduled maintenance visit.

**(b)** checks and adjustments (as appropriate) shall be carried out as detailed below:

    **(i)** ensure that noise is not excessive when the installation is operating above 20 % of maximum capacity.

    **(ii)** with the installation operating at normal load, check and adjust (as necessary) the set point of each regulator. Where fixed factor conversion is applied, ensure that the set pressure is accurate.

    **(iii)** check that the regulator(s) is/are controlling the pressure of the gas at the meter in a stable manner.

    **(iv)** check that the meter index is still operating.

    **(v)** check that the uncorrected volume of gas recorded on any conversion system agrees with the volume registered by the meter to within the tolerance stated by the conversion system manufacturer in IGEM/GM/5.

    **(vi)** any commissioning strainer installed upstream of the meter should be removed. Appropriate procedures should be followed while undertaking this work.

**Note:** This will, typically, require the removal of a spool piece. Thus, there will be a need to isolate the appropriate section of pipe, purge the isolated section to air/inert gas, remove the spool and filter, replace the spool, purge back to gas and undertake an appropriate tightness test.

With the meter and instrumentation operating, equipment and instruments should be carefully observed for signs of faulty operation such as unexplained noise, overheating and effects of vibration.

## B.1.3 Water

### Checklist for water meter installation and commissioning

**(a)** If the water is non-potable (for example, raw or recycled water) and is likely to contain particulate or other debris, an upstream strainer or filter may be required – check manufacturer's recommendations.

© The Institution of Engineering and Technology

**(b)** If the meter is installed outside or in an unheated area consider the need for frost protection.

**(c)** Check that the pressure rating of the meter is sufficient for the line.

**(d)** When cutting pipe to accommodate the meter, ensure that the pipe ends are cleaned and there is no swarf that could jam or damage the meter.

**(e)** Ensure that the meter is fitted with water flowing in the direction of the arrow on the meter body.

**(f)** Ensure that the meter is installed at a location where the pipe will be fully pressurised with no void at the crown of the pipe.

**(g)** Installation in vertical pipes should be such that water rises up through the meter.

**(h)** The pipe immediately upstream and downstream of the meter should be straight, with no fittings, inlets or take-offs and of the same nominal diameter of the meter to at least the length recommended by the meter manufacturer.

**(i)** Isolation valves should be fitted upstream and downstream of the meter to facilitate replacement or maintenance.

**(j)** The meter should be fitted in the orientation recommended by the manufacturer – in a horizontal pipe the register of a mechanical meter should be uppermost.

**(k)** Ensure that the connections are tight with no leaks upstream or downstream.

**(l)** When the pipe is brought into service, bleed air out slowly to avoid overspeeding and damaging the meter.

**(m)** If the water contains substances likely to coat the pipe or meter bore, or lead to sedimentation within the pipe, inspection or rodding points should be included.

**(n)** Large meters should be properly supported to avoid strain on the pipework.

**(o)** Meters and outreaders should be clearly labelled and marked to facilitate identification, particularly where they are in chambers.

## B.1.4 Heat

In order to carry out successful commissioning of the equipment it is required that the following checks are completed and tested. It is strongly recommended that the meter provider carries out a commissioning procedure to confirm that the meter is installed correctly as this may extend the warranty type or period.

**Note:** Always refer to the manufacturer's/supplier's installation documentation.

**(a)** Before commissioning, check that electrical supply, water flow and true condition temperatures will be available at time of commissioning. Commissioning of the metering device cannot be carried out correctly without all of these factors.

**(b)** Confirm that the meter is installed into the correct location/pipework to measure the required service/consumption.

**(c)** Check that all components of the meter are fitted in the system, i.e. the flow meter is fitted in either the flow or the return pipe (preferably in the return pipe) checking for correct flow direction. Temperature sensors and supplied pockets are fitted (one in the high temperature line and one in the low temperature line). Energy integrator is fitted on the wall/via a support bracket in an accessible location.

**(d)** Check that all wiring is complete and marked with identity labels and is terminated correctly as per installation instructions, i.e. power supply cabling fitted (230 V AC, unless stated) to all relevant parts (flow meter and integrator).

**(e)** Check that the internal wiring between all meter elements (flow meter, integrator and temperature sensors) are as shown in installation instruction.

**(f)** Check that the BMS/AMR connection, where applicable, is connected and terminated correctly as per installation instructions. Confirm that this cable has been tested by the BMS/AMR installer prior to connection to the meter.

**(g)** Record all of the meter serial numbers, start readings and include energy, volume, flow rate, flow and return temperatures and power. Check that these meet with the meter criteria and fall within its ranges.

**(h)** The pipe immediately upstream and downstream of the meter should be straight, with no fittings, inlets or take-offs and of the same nominal diameter of the meter to at least the length recommended by the meter manufacturer.

© The Institution of Engineering and Technology

# APPENDIX C

## Glossary

| | |
|---|---|
| ACoP | Approved Code of Practice |
| AMR | Automatic meter reading |
| ATEX | ATmospheres EXplosibles |
| CAD | Consumer access device – a device capable of joining a (British) Smart Metering HAN with access limited to reading metering data (it cannot send data to the meters). It provides a gateway into other networks, for example, the consumer's HAN or an energy manager's WAN. It allows metering data to be collected and used in conjunction with energy management services independent of the energy supplier. |
| COSEM | The Companion Specification for Energy Metering (COSEM) sets the rules for data exchange with energy metering based on IEC 62056 standards. |
| DCC | Data communications company. Awarded the monopoly contract by government to read SMETS meters (SMETS2 exclusively). |
| Dial-in | Opposite to dial-out. Defines which unit initiates the connection (not just telephone but any client/server network). Dial-in means that the meter initiates the call and the interrogating unit is the server. |
| Dial-out | Defines which unit initiates the connection (not just telephone but any client/server network). Dial-out means that the meter is the server and the interrogating unit initiates the call. |
| DLMS | Device language messaging system, maintained by European meter vendors attempting to standardise meter data formats, objects and services. Used mainly for electricity metering but can also be used for the metering of gas, heat and water. The specification defines data objects and methods used in metering in a standardised way to allow meter reading equipment to use standard addressing methods to read and write data to the meters. The IEC 62056 series of standards cover the protocols. |
| DSEAR | Dangerous Substances and Explosive Atmospheres Regulations |
| ECV | Emergency control valve |
| Electrical equipment | Used in conjunction with gas meters, it means that one or more components have been assembled for a specific purpose, including electronic, for example, gas volume conversion devices, dataloggers, and optically-coupled equipment. |
| FTP | File transfer protocol. The universally acknowledged industry standard used to manage files on the internet. |
| GB | Great Britain |
| GBCS | The Great Britain Companion Specification – the DLMS variant for the UK, currently being drafted by BEIS in conjunction with BEAMA and the energy suppliers. |

| | |
|---|---|
| GPRS | General Packet Radio Service. Mobile telephone. Data is transmitted similarly to an internet connection and charged by amount of data. |
| GSM | Global System for Mobile Communications. Mobile telephone. Data is transmitted similarly to a modem on a telephone line and charged by online time. |
| HAN | Home area network. The communications network used independently and privately within the domain of a single consumer premises. |
| HASWA | Health and Safety at Work Act |
| Hazardous area | An area in which explosive gas/air mixtures are, or may be expected to be, present in quantities such as to require special precautions for the construction, installation and use of electrical apparatus or other sources of ignition. BS EN 60079-10 specifies hazardous areas by making reference to three sets of conditions that are recognised and defined:<br><br>Zone 0: an area in which an explosive air/gas mixture is continuously present or is present for long periods.<br><br>Zone 1: an area in which an explosive air/gas mixture is likely to occur in normal operation.<br><br>Zone 2: an area in which an explosive air/gas mixture is not likely to occur in normal operation and, if it occurs, it will exist only for a short time. |
| Head end | Sometimes referred to as a head end system (HES), this is the in-station software that communicates with the meter portfolio. |
| HSE | Health and Safety Executive |
| IGEM | Institution of Gas Engineers and Managers |
| IHD | In-home display. The small display device used to show energy consumption, issued to each consumer as part of the smart meter roll-out. |
| IP | Internet protocol. The universally acknowledged industry standard used to transmit data on private networks or on the internet. |
| LNG | Liquefied natural gas |
| M-Bus | Similar to Modbus, but used primarily in mainland Europe. M-Bus was developed as a two-wire communication system for use with battery powered meters (gas, water, heat). The two wires provide both power and data to minimise battery drain during communications.<br><br>M-Bus is commonly deployed in continental Europe and is covered by the EN 13757 European Standards.<br><br>Wireless M-Bus uses the same protocols over an 868 MHz low power radio link. |
| Metering CTs | Current transformer designed specifically for secondary meter. Protection CTs are built to withstand much higher currents. If a protection CT is used for measuring there is a risk that, should current levels ever rise, the metering equipment may not be able to thermally withstand such high currents. |
| MID | Measuring Instruments Directive. An EU directive for all equipment used to measure consumption for the purposes of billing. |

© The Institution of Engineering and Technology

| | |
|---|---|
| Modbus | A low voltage two-wire signalling network used primarily in the UK to transmit data at low bandwidth between devices and the collecting station or concentrator. |
| Non-hazardous area | A 'safe' area. An area in which an explosive gas/air mixtures is not expected to be present in quantities such as to require precautions for the constant installation and use of apparatus. |
| Ofgem | Office of Gas and Electricity Markets |
| PRI | Pressure regulating installation |
| Protection CT | Current transformer design to provide electrical protection. |
| Regulatory delivery | A Directorate within the Department for Business, Energy and Industrial Strategy and has the statutory responsibility for the metrological performance of gas and electricity meters. |
| SMETS | The Smart Metering System Technical Specification defined by DECC (versions 1 and 2) with which all UK smart meters must comply. |
| SMS | Text messages using mobile phone technology to transmit data packets of up to 160 bytes of metering data per message. |
| USM | Ultrasonic meter. |
| WAN | Wide area network. |
| Wireless M-Bus | A wireless version of M-Bus, using sub GHz frequency. |
| ZigBee | A radio mesh network standard used primarily in the USA, currently being adapted for use as the smart meter HAN in the UK by DECC at 2.4 GHz, with proposals for a sub GHz implementation to provide better coverage. |
| Z-Wave | Z-wave is a radio-based system designed for building control systems using the 868 MHz ISM band. It supports a wide range of control devices and can support the transportation of metering data. |

## Gas Pressure Nomenclature

All pressure settings are on the consumer side, which means that they are downstream of the primary meter installation so that they would be identified by a subscript 'c', i.e. $MOP_c$

| | |
|---|---|
| DMIP | Design maximum incidental pressure – the maximum pressure that a system is permitted to experience under fault conditions, limited by safety devices when the system is operated at the design pressure. |
| DmP | Design minimum pressure – the minimum pressure that may occur at a point, at the time of system design flow rate under extreme gas supply and maintenance conditions. |
| DP | Design pressure – the pressure on which design calculations are based. |
| LOP | Lowest operating pressure – the minimum pressure that a system is designed to experience under normal operating conditions. |
| MIP | Maximum incidental pressure – the maximum pressure that a system is permitted to experience under fault conditions limited by safety devices. |
| MOP | Maximum operating pressure – the maximum pressure under which a system can continuously operate in normal operating conditions. |

© The Institution of Engineering and Technology

| | |
|---|---|
| OP | Operating pressure – the pressure at which a gas system operates under normal conditions. |
| SOL | Safe operating limit – the operating limit (including a margin of safety) beyond which system failure is liable to occur. |
| STP | Strength test pressure – the pressure applied to a system during a strength test. |

© The Institution of Engineering and Technology

# APPENDIX D

## Additional information

### D.1  Summation meters

When the current in a number of feeders need not be individually metered but summated to a single meter or instrument, a summation current transformer can be used.

In consumer installations where there is more than one feeder it is more economical to use summation metering and, for this purpose, a summation CT is required.

The summation CT consists of two or more primary windings that are connected to the feeders to be summated, and a single secondary winding that feeds a current proportional to the summated primary current.

A typical ratio would be 5 + 5 + 5 / 5 A, which means that three primary feeders of 5 A are to be summated to a single 5 A meter. The correct meter ratio must be used for summation metering to take the sum of the feeders.

As an example, if in a two-way summation, each of the CTs are rated at 600 / 5 A, then the meter should have a ratio of 1200 / 5 A in order to give the true reading of both feeders.

If you have multiple circuits to measure (sub-metering) it would be beneficial to introduce summation CTs to the circuit as a whole.

Accuracy again plays an important part because poor quality summation CTs will add up and affect overall accuracy.

▼ **Figure D.1**   Example summation metering schematic

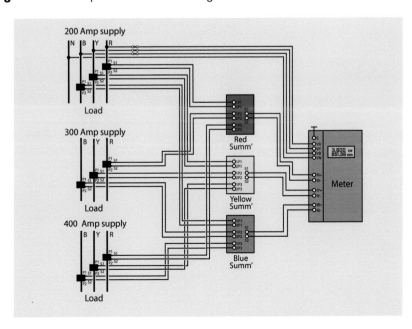

© The Institution of Engineering and Technology

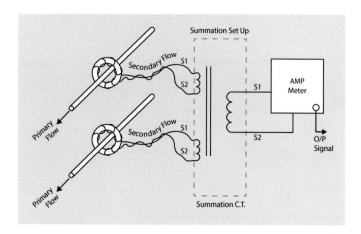

## D.2  Smart metering

### D.2.1 Background

By 2020 there is planned to be over 54 million domestic smart meters installed in the UK for electricity and gas as part of the Smart Meter Implementation Program (SMIP) announced by the government in 2006.

Two phases are planned:

**(a)** Phase 1, where the metering systems meet the Smart Metering System Technical Specification (SMETS) in all but common protocol and security requirements; and

**(b)** Phase 2, or SMETS2, which will include common protocol (GBCS) and enhanced security (SMKI).

SMETS1 meters use different standards for their communications protocols and hence may not be compatible with different suppliers' systems. Consequently, they may have to be changed on change of supplier. SMETS1 meters will cease to be installed when SMETS2 is certified.

© The Institution of Engineering and Technology

▼ **Figure D.3** Smart Meter Implementation Program (SMIP) structure

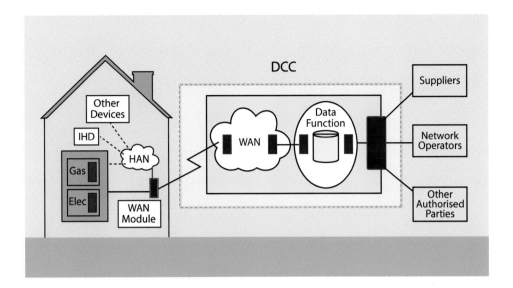

The UK smart metering system comprises of an electricity meter, a gas meter, a communications hub, an in-home display (IHD), a head-end (meter support software) and a communications infrastructure run by the DCC as part of a contract awarded in September 2013. All smart metering activity must be in accordance with the Smart Energy Code (SEC) governed by a panel of industry members, which is accountable to the regulator Ofgem. Changes to metering requirements are requested, evaluated and approved by the SEC Panel.

The WAN uses GPRS in the midlands and southern part of the UK and long range radio in the north (from Manchester).

DLMS/COSEM is the protocol used for electricity messages and a modified form of ZigBee is used for gas messages over the WAN.

DLMS/COSEM messages are tunnelled over ZigBee to the electricity meter on the HAN and gas messages use the ZigBee protocol over the HAN.

There are three ways an end consumer will be able to access metering data. The first source of data is through the IHD. The second is through the DCC – a user will need to register as an acceded SEC "Other User" Party, purchase some adapter software, and pay the DCC for usage on a transactional basis. The third is through a Consumer Access Device (CAD), which can be purchased and installed by the consumer or third party and uses data signals from the HAN.

The HAN is provided through the ZigBee Smart Energy Profile carried over a 2.4 GHz radio link. There may be issues with propagation through some properties depending on their construction and it is envisaged that an 868 MHz based alternative will be available to address these issues. There will be issues in multi-dwelling units (apartment buildings) where the meters may be some considerable distance from the apartment, located in basements. Powerline carrier options are under consideration for these situations.

Smart meters are also used in small industrial premises (or by SMEs) Profile Class 3 and 4, but these are not mandatory for commercial consumers.

© The Institution of Engineering and Technology

# Index

## A

## B

## C

## D

## E

© The Institution of Engineering and Technology

© The Institution of Engineering and Technology

## R

## S

## T

## U

## V

No Entries

© The Institution of Engineering and Technology

## W

## X

No Entries

## Y

No Entries

## Z

IET Standards

# Influence the future
# of Standards

Working in a fast paced, rapidly changing industry has its frustrations. A lack of professional standards and guidance increases risk, and hinders the ability to embrace innovation.

The IET uses its wealth of knowledge and experience to bring about standards that:

- Solve common working problems
- Make meeting legislative requirements simple
- Give practical guidance for practising engineers

To create the best possible guidance, we need you.

Get involved as a member of a publication committee, take on authorship of a book or simply give us feedback on a draft publication. The choice is yours.

**Find out more at:**

**www.theiet.org/setting-standards**

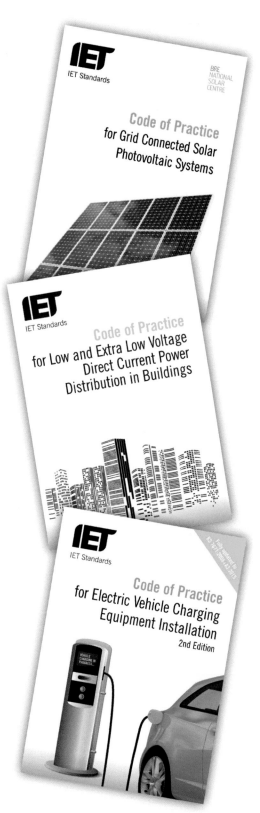

IET Standards

# Industry-leading standards

IET Standards works with industry-leading bodies and experts to publish a range of codes of practice and guidance materials for professional engineers, using its expertise to achieve consensus on best practice in both emerging and established technology fields.

See the full range of IET Standards titles at:

**www.theiet.org/standards**

**Centre of Excellence**
The Institution of Engineering and Technology

# IET Centres of Excellence

The IET recognises training providers who consistently achieve high standards of training delivery for electrical installers and contractors on a range of courses at craft and technician levels.

Using an IET Centre of Excellence to meet your training needs provides you with:

- Courses that have a rigorous external QA process to ensure the best quality training
- Courses that underpin the expertise required of the IET electrical regulations publications
- Training by competent and professional trainers approved by industry experts at the IET

See the current list of IET Centres of Excellence in your area at:

**www.theiet.org/excellence**

Electrical **excellence**

# Expert publications

The IET is co-publisher of BS 7671 (IET Wiring Regulations), the national standard to which all electrical installations should conform. The IET also publishes a range of expert guidance supporting the Wiring Regulations.

You can view our entire range of titles including...

- BS 7671
- Guides
- Guidance Notes series
- Inspection, Testing and Maintenance titles
- City & Guilds textbooks and exam guides

...and more at:

**www.theiet.org/electrical**

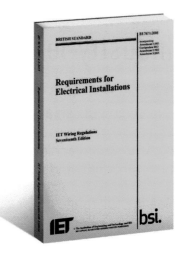

---

## ELECTRICAL STANDARDS

# Constantly up-to-date digital subscriptions

Our expert content is also available through a digital subscription to the IET's Electrical Standards Plus platform. A subscription always provides the newest content, giving peace of mind that you are always working to the latest guidance.

It also lets you spread the cost of updating all your books once new versions are released.

Going digital gives you greater flexibility when working with the Wiring Regulations, Guidance Notes and the IET's expert Codes of Practice available for electrical engineers. The intuitive search function, instantly serves results from across all books in your package. You can also access the content on your desktop, laptop or tablet, making it easy to take the content out on site or read on the move.

Find out more about our subscription packages and choose one to suit you at:

**www.theiet.org/esplus**